Peter Ghavami
Big Data Analytics Methods

Peter Ghavami

Big Data Analytics Methods

Analytics Techniques in Data Mining, Deep Learning and Natural Language Processing

2nd edition

DE GRUYTER

ISBN 978-1-5474-1795-7
e-ISBN (PDF) 978-1-5474-0156-7
e-ISBN (EPUB) 978-1-5474-0158-1

Bibliographic information published by the Deutsche Nationalbibliothek
The Deutsche Nationalbibliothek lists this publication in the Deutsche Nationalbibliografie; detailed bibliographic data are available on the internet at http://dnb.dnb.de.

published by Walter de Gruyter Inc., Boston/Berlin
Cover image: Rick_Jo/iStock/Getty Images Plus
Typesetting: Integra Software Services Pvt. Ltd.
Printing and binding: CPI books GmbH, Leck

www.degruyter.com

To my beautiful wife Massi,
whose unwavering love and support make these accomplishments possible and worth pursuing.

Acknowledgments

This book was only possible as a result of my collaboration with many world renowned data scientists, researchers, CIOs and leading technology innovators who have taught me a tremendous deal about scientific research, innovation and more importantly about the value of collaboration. To all of them I owe a huge debt of gratitude.

Peter Ghavami
March 2019

https://doi.org/10.1515/9781547401567-202

About the Author

Peter Ghavami, Ph.D., is a world renowned consultant and best-selling author of several IT books. He has been consultant and advisor to many Fortune 500 companies around the world on IT strategy, big data analytics, innovation and new technology development. His book on clinical data analytics titled "Clinical Intelligence" has been a best-seller among data analytics books.

His career started as a software engineer, with progressive responsibilities to technology leadership roles such as: director of engineering, chief scientist, VP of engineering and product management at various high technology firms. He has held leadership roles in data analytics including, Group Vice President of data analytics at Gartner and VP of Informatics.

His first book titled *Lean, Agile and Six Sigma IT Management* is still widely used by IT professionals and universities around the world. His books have been selected as text books by several universities. Dr. Ghavami has over 25 years of experience in technology development, IT leadership, data analytics, supercomputing, software engineering and innovation.

Peter K. Ghavami received his BA from Oregon State University in Mathematics with emphasis in Computer Science. He received his M.S. in Engineering Management from Portland State University. He completed his Ph.D. in industrial and systems engineering at the University of Washington, specializing in prognostics, the application of analytics to predict failures in systems.

Dr. Ghavami has been on the advisory board of several analytics companies and is often invited as a lecturer and speaker on this topic. He is a member of IEEE Reliability Society, IEEE Life Sciences Initiative and HIMSS. He can be reached at peter.ghavami@northwestu.edu.

https://doi.org/10.1515/9781547401567-203

Contents

Appendices

Introduction

Data is the fingerprint of creation. And Analytics is the new "Queen of Sciences." There is hardly any human activity, business decision, strategy or physical entity that does not either produce data or involve data analytics to inform it. Data analytics has become core to our endeavors from business to medicine, research, management, product development, to all facets of life.

From a business perspective, *data* is now viewed as the new *gold*. And *data analytics*, the machinery that mines, molds and mints it. Data analytics is a set of computer-enabled analytics methods, processes and discipline of extracting and transforming raw data into meaningful insight, new discovery and knowledge that helps make more effective decisions. Another definition describes it as the discipline of extracting and analyzing data to deliver new insight about the past performance, current operations and prediction of future events.

Data analytics is gaining significant prominence not just for improving business outcomes or operational processes; it certainly is the new tool to improve quality, reduce costs and improve customer satisfaction. But, it's fast becoming a necessity for operational, administrative and even legal reasons.

We can trace the first use of data analytics to the early 1850s, to a celebrated English social reformer, statistician and founder of modern nursing, Florence Nightingale.[1] She has gained prominence for her bravery and caring during the Crimean War, tending to wounded soldiers. But her contributions to statistics and use of statistics to improve healthcare were just as impressive. She was the first to use statistical methods and reasoning to prove better hygiene reduces wound infections and consequently soldier fatalities.

At some point during the Crimean War, her advocacy for better hygiene reduced the number of fatalities due to infections by 10X. She was a prodigy who helped popularize graphical representation of statistical data and is attributed to have invented a form of pie-chart that we now call *polar area diagram*. She is attributed with saying: "To understand God's thoughts we must study statistics, for these are the measure of his purpose." Florence Nightingale is arguably the first data scientist in history.

Data analytics has come a long way since then and is now gaining popularity thanks to eruption of five new technologies called SMAC: social media, mobility, analytics, and cloud computing. You might add another to the acronym for sensors, and the internet of things (IoT). Each of these technologies is significant in how they transform the business and the amount of data that they generate.

---

1 Biography.com, http://www.biography.com/people/florence-nightingale-9423539, accessed December 30, 2012.

https://doi.org/10.1515/9781547401567-001

Portrait of Florence Nightingale, the First Data Scientist

In 2001, META (now Gartner) reported a substantial increase in the size of data, the increasing rate at which data is produced and wide range of formats. They termed this shift *big data*. Big data is known by its three key attributes, known as the three V's: volume, velocity, and variety. Though, four more V's are often added to the list: veracity, variability, value and visualization.

The world storage volume is increasing at a rapid pace, estimated to double every year. The velocity at which this data is generated is rising, fueled by the advent of mobile devices and social networking. In medicine and healthcare, the cost and size of sensors has shrunk, making continuous patient monitoring and data acquisition from a multitude of human physiological systems an accepted practice. The internet of things (IoT) will use smart devices that interact with each other generating the vast majority of data, known as machine data, in the near future.

Currently 90% of big data is known to have accumulated in the last two years. Pundits estimate that by 2020, we will have 50 times the amount of data we had in 2011. It's expected that self-driving cars will generate 2 Petabytes of data every year. Cisco predicts that by 2022 the mobile data traffic will reach 1 zettabyte.[2] Another article puts the annual growth of data at 27% per year, reaching 333 exabytes per month by 2022.[3]

With the advent of smaller, inexpensive sensors and volume of data collected from customers, smart devices and applications, we're challenged with making increasingly analytical decisions from a large set of data that are being collected in the moment. This trend is only increasing giving rise to what's known in the

2 Article by Wie Shi, "Almost One Zettabyte of Mobile Data Traffic in 2022," published by Telecoms.com.

3 Statista.com article, "Data Volume of Global Consumer IP Traffic from 2017 to 2022."

industry as the "big data problem": The rate of data accumulation is rising faster than our cognitive capacity to analyze increasingly large data sets to make decisions. The big data problem offers an opportunity for improved predictive analytics and prognostics.

The variety of data is also increasing. The adoption of digital transformations across all industries and businesses is generating large volume and diverse data sets. Consider the medical data that was confined to paper for too long. As governments such as the United States push medical institutions to transform their practice into electronic and digital format, patient data can take diverse forms. It's now common to think of electronic medical record (EMR) to include diverse forms of data such as audio recordings, MRI, ultrasound, computed tomography (CT) and other diagnostic images, videos captured during surgery or directly from patients, color images of burns and wounds, digital images of dental x-rays, waveforms of brain scans, electro cardiogram (EKG), genetic sequence information and the list goes on.

IDC[4] predicted that the worldwide volume of data would increase by 50X from 2010 to 2020. The world volume of data will soon reach 44ZB (zettabytes) by 2020.[5] By that time, new information generated for every human being per second will be around 1.7 megabytes.[6] Table I.1 offers a relative sizing of different storage units of measure.

Table I.1: Storage units of measure.

Data Volume	Size
Bytes – 8 Bits	1 byte: a single character
Kilobyte – 1000 Bytes	A very short story
Megabyte – 1000 KiloBytes	A small novel
Gigabyte – 1000 MegaBytes	A movie at TV quality
Terabyte – 1000 GigaBytes	All X-ray films in a large hospital
Petabyte – 1000 TeraBytes	Half of all US academic research libraries
Exabyte – 1000 PetaBytes	Data generated from SKA telescope in a day
Zettabyte – 1000 ExaBytes	All worldwide data generated in 1st half of 2012
Yottabyte – 1000 ZetaBytes	1 YB = 1000^8 bytes – 10^{24} byes

4 International Data Corporation (IDC) is a premier provider of research, analysis, advisory and market intelligence services.

5 Each Zettabyte is roughly 1000 Exabytes and each Exabyte is roughly 1000 Petabytes. A Petabyte is about 1000 TeraBytes.

6 https://www.newgenapps.com/blog/big-data-statistics-predictions-on-the-future-of-big-data

The notion of all devices and appliances generating data has led to the idea of the internet of things, where all devices communicate freely with each other and to other applications through the internet. McKinsey & Company predicts that by 2020, big data will be one of the five game changers in US economy and one-third of the world data will be generated in the US.

New types of data will include structured and unstructured text. It will include server logs and other machine generated data. It will include data from sensors, smart pumps, ventilators and physiological monitors. It will include streaming data and customer sentiment data about you. It includes social media data including Twitter, Facebook and local RSS feeds about healthcare. Even today if you're a healthcare provider, you must have observed that your patients are tweeting from the bedside. All these varieties of data types can be harnessed to provide a more complete picture of what is happening in delivery of healthcare.

Big data analytics is finding its own rightful platform at the corporate executive C-suite. Job postings for the role of Chief Data Officer are rapidly growing. Traditional database systems were designed to handle transactions rapidly but not designed to process and handle large volumes, velocity and variety of data. Nor are they intended to handle complex analytics operations such as anomaly detection, finding patterns in data, machine learning, building complex algorithms or predictive modeling.

The traditional data warehouse strategies based on relational databases suffer from a latency of up to 24 hours. These data warehouses can't scale quickly with large data growth and because they impose relational and data normalization constraints, their use is limited. In addition, they provide retrospective insight and not real-time or predictive analytics.

The value proposition of big data analytics in your organization is derived from the improvements and balance between cost, operations and revenue growth. Data analytics can identify opportunities to grow sales and reduce costs of manufacturing, logistics and operations. The use cases under these three categories are enormous. It can also aid in cyber security and big data analysis.

Deriving value from data is now the biggest opportunity and challenge for many organizations. CEOs ask how do we monetize the data that we have in our databases? Often the answer includes not just analyzing internal data but combining with data from external sources. Crafting the data strategy and use-cases is the key to leveraging huge value from your data.

According to a McKinsey & Company research paper, big data analytics is the platform to deliver five values to healthcare: Right living, Right Care, Right Provider, Right Value, Right Innovation.[7] These new data analytics value systems

7 "The Big Data Revolution in Healthcare," Center for US Health System Reform, McKinsey & Co. (2013).

drive boundless opportunities in improving patient care and population health on one hand and reducing waste and costs on the other.

In many domains and industries, for example the medical data, we're not just challenged by the 3 V's. Domain specific data brings its own unique set of challenges that I call the 4 S's: Situation, Scale, Semantics and Sequence. Let's evaluate the 4S categories in the context of medical data.[8]

Taking data measurements from patients has different connotation in different situations. For example, a blood pressure value taken from a patient conveys a different signal to a doctor if the measurement was taken at rest, while standing up or just after climbing some stairs. Scale is a challenge in medicine since certain measurements can vary drastically and yet remain clinically insignificant compared to other measurements that have a limited rate of change but a slight change can be significant.

Some clinical variables have a limited range versus others that have a wider range. For example; analyzing data that contains patient blood pressure and body temperature that have a limited range requires understanding of scale since a slight change can be significant in the analysis of patient outcome. In contrast, a similar amount of fluctuation in patient fluids measured in milliliters may not be serious. As another example, consider the blood reticulocytes value (rate of red blood cell production). A normal Reticulocyte value should be zero, but a 1% increase is cause for alarm, an indication of body compensating for red blood cell count, a possible body compensation to shock to the bone marrow.

We can see the complexity associated with scale and soft thresholds best in lab blood tests: the normal hemoglobin level in adults is somewhere between 12 to 15 and a drop to 10 a physician might choose to prescribe iron supplements but measured at 4 will require a blood transfusion.

Semantics is critical to understanding data and analytics results. As much as 80% of data is non-structured data in form of narrative text or audio recording. Correctly extracting the pertinent terms from such data is a challenge. Tools such as natural language processing (NLP) methods combined with ontologies and domain expert libraries are used to extract useful data from patient medical records. Understanding sentence structure and relationships between terms are critical to detecting customer sentiment, language translation and text mining.

That brings us to the next challenge in data analytics: sequence. Many activities generate time series data; that is data from activities that occur sequentially. Times series data can be analyzed using different techniques such as ARIMA (Autoregressive Integrated Moving Average) and Markov Chains. Keeping and analyzing data in its sequence is important. For example, physiological and clinical data collected over certain time periods, during a patient's hospital stay can be studied as sequential or time series

8 *Clinical Intelligence: The Big Data Analytics Revolution in Healthcare – A Framework for Clinical and Business Intelligence*, Peter Ghavami (2014).

data. The values measured at different times are significant and in particular the sequence of those values can have different clinical interpretations.

In response to these challenges, big data analytics techniques are rapidly emerging and converging making data analytics more practical and common place. One telling sign is the growing interest and enthusiasm in analytics competitions, analytics social groups, and meets ups which are growing rapidly.

Kaggle, a company that hosts open machine learning competitions, started in 2011 with just one open competition. At the time of writing of this book, Kaggle[9] is hosting hundreds of competitions in a year. In 2016, it had received more than 20,000 models from data scientists around the globe competing for the highest ranking. In 2018, the number of submissions had reached 181,000. A few interesting competitions included the following companies and problems:

Merck: The Company offered a price of $100,000 to the best machine learning model that could answer one question. Given all our data on drug research which chemical compounds will make good drugs?

Genentech: A member of the Roche Group, Genentech offered $100,000 to the best classification and predictive program for cervical cancer screening to identify which individuals from a population are at risk of cervical cancer.

Prudential Insurance: The Company wants to make buying life insurance easier. It offers a $30,000 prize for the best predictive model that determines which factors are predictors for households to buy life insurance.

Airbnb: The Company started an open competition for the best predictive model that predicts where customers are likely to book their next travel experience.

Later in the book, we'll define what "best model" means, as the word "best" can mean many things. In general we use the best model based on improved performance, accuracy of prediction, robustness of the model against diverse data sets and perhaps speed of learning; all these go together into defining the best attributes of an analytics model.

Over the last decade, I've been asked many questions by clients that share common threads. These are typical questions nagging data scientists and business leaders alike. I frequently hear questions like: What is machine learning? What is the difference between classification and clustering? How do you clean dirty and noisy data? How do you handle missing data? And so on. I've compiled answers to these questions in this book to provide guidance to those who are passionate about data analytics as I am.

Other leading companies use data analytics to predict their customer's needs. Retailers such as Target are able to predict when customers are ready to make a purchase. Airbnb predicts when a client is likely to take a vacation and conducts

9 www.kaggle.com

targeted marketing to make a pitch for a specific get away plan that appeals to the customer. Use of smartphones as a platform to push in-the-moment purchases has become a competitive advantage for several companies.

Other companies are pushing messages to their client mobile devices to invite them to their stores offering deals at the right time and the right place. Several institutions have improved their revenues by predicting when people are likely to shop for new cars. One institution uses machine learning and combination of behavioral data (such as online searches) to predict when a customer is likely to purchase a new car and offers tailored car packages to customers.

This book is divided into three parts. Part I covers the basics of analytics, topics like correlation analysis, multivariate analysis and traditional statistical methods. Part II is concerned with advanced analytics methods, including machine learning, classifiers, cluster analysis, optimization, predictive modeling and Natural Language processing (NLP). Part III includes a case study to illustrate predictive modeling, validation, accuracy and details about ensemble of models.

Prediction has many important use-cases. Predicting consumer behavior provides the opportunity to present in-the-moment deals and offers. Predicting a person's health status can prevent escalating medical costs. Predicting patient health condition provides the opportunity to apply preventive measures that result in better patient safety, quality of care and lower medical costs; in short, timely prediction can save lives and avoid further medical complications. Predictive methods using machine learning tools such as artificial neural networks (ANN) promise to deliver new intuitions into the future; giving us insight to avert a disaster or seize an opportunity.

Advances in software, hardware, sensor technology, miniaturization, wireless technology and mass storage allow recording and analysis of large amounts of data in a timely fashion. This provides both a challenge and an opportunity. The challenge is that the decision maker must sift through vast amount of data, fast and complex data, to make the appropriate business decision. The opportunity is to analyze this large amount of fast data in real time to provide forecasts about individual's needs and assist with the right solutions.

A survey conducted by Aberdeen Group revealed that the best-in-class healthcare organizations (those who rank higher on the key performance indicators), were much more savvy and familiar with data analytics than the lower performing healthcare organizations.[10] In fact, 67% of the best-in-class providers used clinical analytics, versus only 42% analytics adoption among the low-performing providers. In terms of ability to improve quality, the best-in-class providers using analytics were twice as capable (almost 60% vs. 30%) as the low-performing providers to respond and resolve quality issues. One take away from this research was that healthcare providers who don't use

10 "Healthcare Analytics: Has the Need Ever Been Greater?" By David White, Aberdeen Group, A Harte-Hanks Company, September 2012.

analytics are unable to make the proper process and quality changes because they are simply not aware of the relevance and needed facts and metrics to make those changes.

Other studies have demonstrated similar results. Big data analytics promises phenomenal improvements to organizations in any industry. But, as an IT investment we should gauge return on investment (ROI) and define other criteria for success.

A successful project must demonstrate palpable benefits and value derived from new insights. Implementing big data analytics will ultimately become a necessary and standard procedure for many organizations as they strive to identify any remaining opportunities in improving efficiency, raising revenues and cutting costs.

When you work in data analytics long enough, you'll discover better techniques to adopt, some best practices to follow and some pitfalls to avoid. I've compiled a short list to share with you. With this broad introduction, I'd like to condition the book to convey several key lessons that I'll introduce as Ghavami's 8 Laws of Analytics. So, what are the 8 laws of analytics? Here is the complete list:

Ghavami's 8 Laws of Analytics

1. **More data is better** – More data means more insight, more intelligence and better machine learning results.
2. **Even small amount of data can be sufficient** – You can use extrapolation techniques on small amount of data to generalize insights for a larger population.
3. **Dirty & Noisy data can be cleaned with analytics** – You can compensate for dirty and noisy data with analytics. There are analytics models that can overcome these issues.
4. **Distinguish signal from noise by signal boosting** – You can boost the effect of signal in your data to overcome noise or presence of too many data variables.
5. **Regularly retrain Machine Learning models as they atrophy over time** – You must regularly retrain your models as Machine learning models lose their accuracy over time.
6. **Be leery of models that are highly accurate** – Never claim a model is 100% accurate or even 95% accurate. A model that is so accurate is likely to be over trained and over fitted to the specific data and hence performs poorly on other data sets.
7. **Handle uncertainty in data not sensitivity** – Data changes drastically over time and from one situation to another. Make your models robust enough to handle a variety of data and data variations.
8. **Ensemble of models improves accuracy** – Use ensemble of models to improve accuracy in prediction, classification and optimization, since multiple models compensate each other's limitations.

These are among holistic benefits of data analytics that produce exceptional return on investment; what may be termed as return on data (ROD) or return on analytics (ROA).

The future of data analytics in the world of tomorrow is bright and will continue to shine for many years to come. We're finally able to shed light on the data that has been locked up in the darkness of our electronic systems and data warehouses for years. When you consider many diverse applications of data analytics ranging from genomics to internet of things (IoT), there are endless opportunities to improve business outcomes, reduce costs, and improve people's life experience using data analytics.

This book was written to provide an overview of data science methods, to be a reference guide for data scientists and provide a body of knowledge for data analytics. While the field of data analytics is changing rapidly, every effort is made to make this book up to date and relevant, for those who are experienced data scientists or for beginners who want to enter this profession.

Now, let's start our journey through the book.

Part I: **Big Data Analytics**

Chapter 1
Data Analytics Overview

1.1 Data Analytics Definition

Data Analytics should be contrasted from business intelligence for two reasons: First, business intelligence (BI) deals with raw business data, typically structured data, and provides insight and information for business decision making. It is used and defined broadly to include business data query and analysis. In contrast data analytics deals with deep insights from the data that go beyond the internal data including external data, diverse data formats and data types, unstructured as well as structured data. Data analytics utilizes more advanced statistical methods and analytics modeling than BI and often deals with much more complex and unstructured data types.

Data analytics increasingly deals with vast amount of data—mostly unstructured information stored in a wide variety of mediums and formats—and complex data sets collected through fragmented databases during the course of time. It deals with streaming data, coming at you faster than traditional RDBMS systems can handle. This is also called fast data. It's about combining external data with internal data, integrating it and analyzing all data sets together.

Data analytics approaches data schema from a different angle. BI analysis deals with structured data mostly stored in RDBMS systems which treat data schema on write. This implies that we must define the data schema before storing the data in a data warehouse. But, big data analytics deals with data schema on read, programmatically by the data engineer or data scientist as part of preparing data for analysis.

When using this broad definition, data analytics requires data collection, data integration, data transformation, analytical methods, decision support, business rules, reporting and dashboards. A broader definition would add data management, data quality, and data warehousing to the mix. Higher adoption of electronic medical records and digital economy are creating a big data opportunity, making big data analytics more relevant and feasible.

There are similar challenges yet significant differences between data analytics and business intelligence. Many of the challenges to get the right business intelligence (BI) are the same in getting the right data analytics. Business intelligence has been defined as the ability to understand the relationships of presented facts in such a way to guide action towards a desired goal.[11]

This definition could apply to both BI and data analytics. But on closer examination, their differences are critical to note.

11 Hans Peter Luhn, "A Business Intelligence System," *IBM Journal*, Vol. 2, No. 4, 314, 1958.

https://doi.org/10.1515/9781547401567-002

One difference is the nature of data and the other is purpose. Business intelligence provides business insight from raw data for the purpose of enabling strategy, tactics, and business decision making. In contrast big data analytics strives to provide insight to enable business decisions from vast amounts of data which are often ambiguous, incomplete, conditional and inconclusive. The third difference is that often higher accuracy of analysis is needed to make the right decisions. These factors combine to create a complex analytical environment for the data scientists and data analytics practitioners.

Big data analytics aims to answer three domains of questions. These questions explain what has happened in the past, what is happening right now and what is about to happen.

The retrospective analytics can explain and present knowledge about the events of the past, show trends and help find root-causes for those events. The real-time analysis shows what is happening right now. It works to present situational awareness, alarms when data reaches certain threshold or send reminders when a certain rule is satisfied. The prospective analysis presents a view in to the future. It attempts to predict what will happen, what are the future values of certain variables. Figure 1.1 shows the taxonomy of the three analytics questions.

The Past	The Present	The Future
Retrospective View	**Real-time View**	**Prospective View**
– What happened?	– What is happening now?	– What will happen next?
– Why it happened?	– Uses real-time data	– How can I intervene?
– Uses historical data	– Actionable dashboards	– Uses historical and real-time data
– Delivers static dashboards	– Alerts	– Predictive dashboards
	– Reminders	– Knowledge-based dashboards

Figure 1.1: The three temporal questions in big data analytics.

1.2 The Distinction between BI and Analytics

The purpose of business intelligence (BI) is to transform raw data into information, insight and meaning for business purposes. Analytics is for discovery, knowledge creating, assertion and communication of patterns, associations, classifications and learning from data. While both approaches crunch data and use computers and software to do that, the similarities end there.

With BI, we're providing a snapshot of the information, using static dashboards. We're working with normalized and complete data typically arranged in rows and columns. The data is structured and assumed to be accurate. Often, data that is out of range or outlier are removed before processing. Data processing uses simple, descriptive statistics such as mean, mode and possibly trend lines and simple data projections to extrapolation about the future.

In contrast data analytics deals with all types of data both structured and unstructured. In medicine about 80% of data is unstructured and in form of medical notes, charts and reports. Big data analytics approaches do not mandate data to be clean and normalized. In fact, they make no assumption about data normalization.

Data analytics may analyze many varieties of data to provide views into patterns and insights that are not humanly possible. Analytics methods are dynamic and provide dynamic and adaptive dashboards. They use advanced statistics, artificial intelligence techniques, machine learning, deep learning, feedback and natural language processing (NLP) to mine through the data. They detect patterns in data to provide new discovery and knowledge. The patterns have a geometric shape and these shapes as some data scientists believe, have mathematical representations that explain the relationships and associations between data elements.

Unlike BI dashboards that are static and give snapshots of data, big data analytics methods provide data exploration, visualization and adaptive models that are robust and immune to changes in data. The machine learning feature of advanced analytics models is able to learn from changes in data and adapt the model over time. While BI uses simple mathematical and descriptive statistics, big data analytics is highly model-based. A data scientist builds models from data to show patterns and actionable insight. Feedback and machine learning are concepts found in data analytics not in BI. Table 1.1 illustrates the distinctions between BI and data analytics.

Table 1.1: The differences between business intelligence and data analytics.

Business Intelligence	Data Analytics
Information from processing raw data	Discovery, insight, patterns, learning from data
Structured data	Unstructured & structured data
Simple descriptive statistics	NLP, classifiers, machine learning, pattern recognition, predictive modeling, optimization, model-based
Tabular, cleansed & complete data	Dirty data, missing & noisy data, non-normalized data
Normalized data	Non-normalized data, many types of data elements
Data snapshots, static queries	Streaming data, continuous updates of data & models, feedback & auto-learning
dashboards snapshots & reports	Visualization, knowledge discovery

1.3 Why Advanced Data Analytics?

For years, the most common and traditional form of data analysis has been grounded in linear and descriptive analytics mostly driven by the need for reporting key performance measures, hypothesis testing, correlation analysis, forecasting and simple statistics; no artificial intelligence was involved.

But, big data analysis goes beyond descriptive statistics. While descriptive statistics are important to understanding and gaining insight about data, big data analysis covers broader and deeper methods to study data and interpret the results. These methods include machine learning (ML), predictive, classification, semantic analysis and non-linear algorithms and as well as the introduction of multi-algorithm approaches.

Traditionally, descriptive statistics answer "what" but offer little help on "why" and "how." They are good at making generalizations about one population versus another, but perform poorly on an individual basis. One example of analytics is classification. A descriptive statistics measure might suggest that 65% of patients with certain preconditions to a disease respond to a specific therapy. But, when a patient is diagnosed with the disease how can we determine if the patient is among the 65% of the population?

Descriptive statistics look at the past events, but it's not ideal for predicting what will happen in the future. Similarly, descriptive statistics offer little insight about causal relationships that help researchers identify root causes of input variables that produce an outcome. While descriptive analytics offers simple tools to determine what is happening in the environment of care, and populations of patients, they come short in giving us the details often necessary to make more intelligent and dynamically adaptive decisions. Big data analytics emphasizes building models and uses model building as a repeatable methodology for data analysis.

Big data analysis can help with customer classifications not just by the traditional demographics factors such as age, gender and life styles, but by other relevant characteristics related to a diverse set of data collected from primary and secondary sources including sources of data exhaust. The definitions for primary, secondary and exhaust data are fuzzy. But, here is an example to illustrate. When you make an electronic purchase on a mobile device your transaction produces primary data. Secondary data might include the geolocation of your purchase. Data exhaust is the side effect of the transaction. For example, the amount of time you took to complete the transaction.

Big data analysis gives us the ability to perform multi-factorial analysis to determine the utility (or value) associated with different courses of strategy and execution factors. Such analysis reveals the key indicators, predictors and markers for observed outcomes. Analytics enables us to "see" these indicators including previously over-looked indicators and apply the correct weight (or consideration) to these indicators when making decisions.

Big data analysis can be used to calculate more accurate and real time measure of business risk, predictors of business outcomes and customer's next move. It can analyze structured and unstructured data to deliver quantitative and qualitative analysis. It can learn from markets, customer data and recommend best options for any given situation.

However, there are many challenges related to the format, meaning and scale of data. To compound the problem, much of the data is unstructured, in form of free text in reports, charts and even scanned documents. There is a lack of enterprise wide dictionary of data terms, units of measure and frequency of reporting. Much of the big data may have data "quality" issues: data may be missing, duplicate, sparse and just not specific enough for a particular type of study. I'll show strategies to overcome the data quality issues in this book.

Going forward, the big data analysis tools will perform 3D's of data analytics as their core tasks: discover, detect, and distribute. The leading big data analytical solution will discover data across disparate and fragmented datasets that reside in various medical, administrative and financial systems. Second, it can aggregate such data—often in real time—normalize and index data on demand. Then perform analytics on the data including semantic analysis through natural language processing (NLP). Finally, it must be able to distribute some actionable insights to decision makers and users (mobile devices carried by physicians or other care providers).

1.4 Analytics Platform Framework

When considering building analytics solutions, defining data analytics strategy and governance are recommended. One of the strategies is to avoid implementing point-solutions that are stand-alone applications which do not integrate with other analytics applications. Consider implementing an analytics platform that supports many analytics applications and tools integrated in the platform. A 4-layer framework is proposed here as the foundation that supports the entire portfolio of analytics and data science applications across the enterprise. The 4-layer framework consists of a data management layer, an analytics engine layer and a presentation layer as shown in Figure 1.2.

In practice, you'll make choices about what software and vendors to adopt for building this framework. The layers are color coded to match a data bus architecture shown in Figure 1.3. The data management layer includes the distributed or centralized data repository. This framework assumes that the modern enterprise data warehouses will consist of distributed and networked data warehouses.

The analytics layer may be implemented using SAS, Python and R statistical language or solutions from other vendors who provide the analytics engines in this layer.

The presentation layer may consist of various visualization tools such as Tableau, QlikView, SAP Lumira, Hitachi Pentaho or McKesson SpotFire, Microsoft

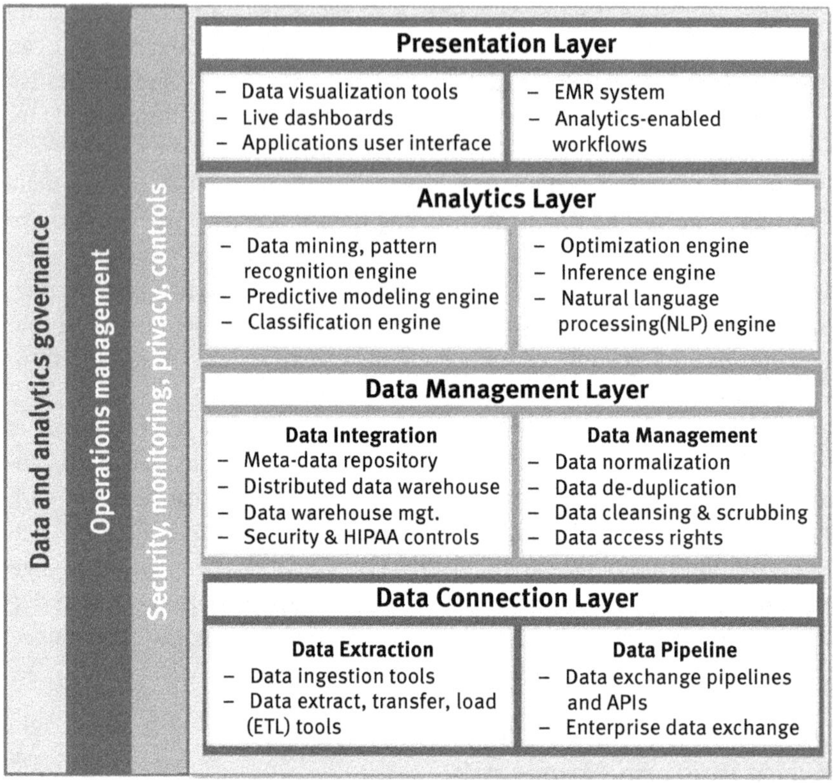

Figure 1.2: The 4-layer data analytics framework.

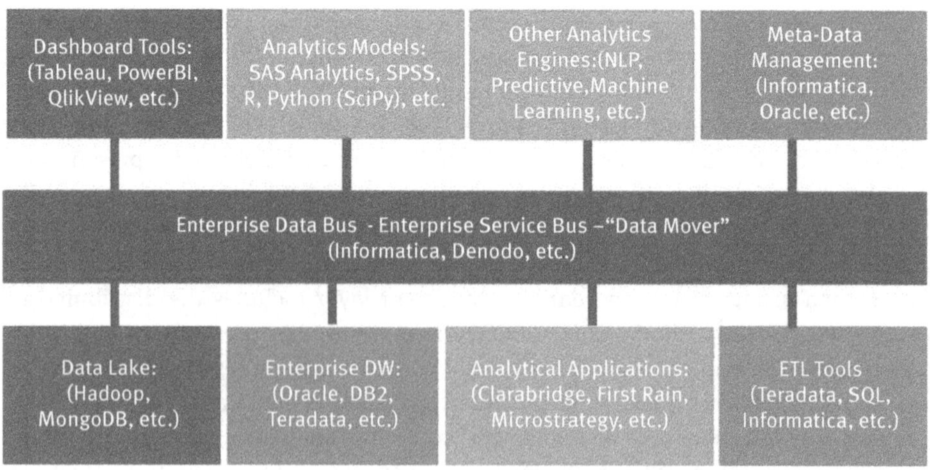

Figure 1.3: The Enterprise Data Bus architecture schematic with example vendor solutions.

PowerBI, Amazon Quicksight and other applications including a data analytics dashboard and application.

In a proper implementation of this framework, the user-facing business application offers analytics-driven workflows and therefore tight integration between the business system and the other two layers (data and analytics) are critical to successful implementation. In many organizations, a data bus architecture may be preferred. This is a highly distributed and federated data strategy. The data bus architecture moves data between various data warehouses, sources of data and data destination (such as the data lake[12]). In addition, the data analytics *engines* shown are modular, *use-case specific analytics programs* that analyze any segment and size of a data set transported via the data bus. These analytics engines can reside in a Hadoop data lake or on a stand-alone system.

The "data mover" component is often referred to as the enterprise service bus (ESB). ESB is defined as a software component that handles communication between mutually interacting software applications in a service-oriented architecture (SOA). ESB facilitates communication between applications, data warehouses and the data lake.

Some notes and explanations for Figure 1.3 are necessary. The references to vendor and product names are by no means an endorsement of these products. The list of companies includes the usual database vendors: IBM, Oracle, Informatica, SAP and Microsoft. Cloud based solutions such as Amazon Glue and Microsoft Data Factory are tools that can fulfill the ESB function in this architecture. These are only provided for examples. I encourage you to perform your own research and comparative analysis to architect a data analytics infrastructure that fits your situation.

The analytics engines referenced here include NLP (natural language processing) and ML (machine learning engine), but there are other engines that the organization can obtain or build to perform specific, targeted analytics functions.

1.5 Data Connection Layer

In the data connection layer, data analysts set up data ingestion pipelines and data connectors to access data. They might apply methods to identify metadata in all source data repositories. Building this layer starts with making an inventory of where the data is created and stored. The data analysts might implement extract, transfer, and load (ETL) software tools to extract data from their source. Other data exchange standards such as X.12 might be used to transfer data to the data management layer.

---

12 As we'll review in later chapters, a data lake is a data storage facility that data streams bring data to (hence the name). Data lake is typically a non-structured storage area for all types of data including internal, external, structured or unstructured data.

In some architectures, the enterprise data warehouse may be connected to data sources through data gateways, data harvesters and connectors using APIs. Products offered by Informatica, Amazon AWS, Microsoft Data Factory, and Talend or similar systems are used as data connector tools.

1.6 Data Management Layer

Once the data has been extracted, data scientists must perform a number of functions that are grouped under the data management layer. The data may need to be normalized and stored in certain database architectures to improve data query and access by the analytics layer. We'll cover taxonomies of database tools including SQL, NoSQL, Hadoop, Spark and other architecture in the upcoming sections.

In the data management layer, we must pay attention to data governance, data security and privacy. We're required to observe HIPAA standards for security and privacy. Jurisdictional regulations and data sovereignty laws will limit transfer of data from one data source to another or from one data center in one country to another.

The data scientist must overcome these limitations with innovative data modeling and data transformation techniques. They will use the tools in this layer to apply security controls, such as those from HITRUST (Health Information Trust Alliance). HITRUST offers a Common Security Framework (CSF) that aligns HIPAA security controls with other security standards.

Data scientists may apply other data cleansing programs in this layer. They might write tools to de-duplicate (remove duplicate records) and resolve any data inconsistencies. Once the data has been ingested, it's ready to be analyzed by engines in the next layer.

Since big data requires fast retrieval, several organizations, in particular the various open source foundations have developed alternate database architectures that allow parallel execution of queries, read, write and data management.

There are three architectural taxonomies or strategies for storing big data that impact data governance, management and analytics:

1. Analyze Data in-Place: Traditionally, data analysts have used the native application and SQL query the application's data without moving the data. Many data analysts' systems build analytics solutions on top of an application's database without using data warehouses. They perform analytics in place, from the existing application's data tables without aggregating data into a central repository. The analytics that are offered by EMR (electronic medical records) companies as integrated solutions to their EMR system fit this category.
2. Build Data Repository: Another strategy is to build data warehouses to store all the enterprise data in a central repository. These central repositories are often known as enterprise data warehouses (EDW). Data from business systems,

customer relationship management (CRM) systems, enterprise resource planning (ERP) systems, data warehouses, financial, transactional and operational systems are normalized and stored in these data warehouses. A second approach called data lake has emerged. Data lakes are often implemented using Hadoop distributed file system or through cloud storage solutions.

The data is either collected through ETL extraction (batch files) or via interface programs and APIs. Data warehouses have four limitations: they often lag behind the real time data by as much as 24 hours; they apply relational database constraints to data which adds to the complexity of data normalization; their support for diverse, new data types is nascent; and they're difficult to use and slow to handle data analytics search and computations. Variations of this architecture include parallel data warehouses (PDW) and distributed data warehouses (DDW).

3. Pull Data on-Demand: An alternate approach is to build an on-demand data pull. This schema leaves the data in the original form (in the application) and only pulls the data when it's needed for analysis. This approach, adopted by only a few hospitals, utilizes an external database that maintains pointers to where the data resides in the source system. The external database keeps track of all data and data dictionaries. When an analytics application requires a piece of data, the external databases pulls the data from the source system on demand and discards it when done. There are two distinct approaches to accomplish this architecture. One is through the use of the Enterprise Service Bus. The other approach is through data virtualization. In data virtualization, an application can access the data regardless of where the data is stored. These approaches have not been widely adopted because they are both difficult to implement and the technology needs to improve.

Of the three options, the data warehousing approach offered tremendous promise, but has proven to be expensive, fraught with time consuming efforts to write ETL programs and develop relational schema to store the data. Pulling data on-demand accelerates access to data since it does not require developing time-consuming data schema development.

Analytics-driven organizations have adopted a two-pronged data strategy that consists of the data warehouse and data lake technologies. Organizations realized that data warehouse is ideal for proven, enterprise class and widely consumed reports. While data lake is ideal for rapid data preparation for data science and advanced analytics applications.

Most analytics models require access to the entire data set because often they annotate the data with tags and additional information which are necessary for models to perform. However, modern architectures prescribe a federated data warehouse model using an enterprise service bus (ESB) or data virtualization. Figure 1.3 illustrates the federated data network model for all sources of data. The data bus connects the sources of data to the analytics tools and to the user-facing applications and dashboards.

The enterprise service bus architecture in Figure 1.3 depicts a federated, networked data architecture scenario for an enterprise. Any data source can also be a data consumer. The primary data sources are application systems, along with the dashboard tools that serve as presentation layer components. The analytics tools in a typical scenario are provided by SAS, and by R Statistical language, but other tools and platforms are acceptable solutions. For example, MicroStrategy can serve as a data platform as well as a data analytics tool. Your data bus architecture components will vary depending on your pre-exiting infrastructure and legacy data warehouse components.

The traditional data management approaches have adopted a centralized data warehouse approach. Despite enormous investments of money and time, the traditional data warehouse approach of building a central repository, or a collection of data from disparate applications has not been as successful as expected. One reason is that data warehouses are expensive and time consuming to build. The other reason is that they are often limited to structured data types and difficult to use for data analytics, in particular when unstructured data is involved. Finally, traditional data warehouses insist on a relational data schema, either dimensional or tabular structures that require the data to meet certain relational integrity or be normalized. Such architectures are not fast enough to handle the large volume of data queries required by data analytics models.

***Star Schema vs. Snowflake Schema*:** The more advanced data warehouses have adopted Kimball's star schema or snowflake schema to overcome the normalization constraints. The star schema splits the business process into fact tables and dimension tables. Fact tables describe measurable information about entities while dimension tables store attributes about entities. The star schema contains one or more fact tables that reference any number of dimension tables. The logical model typically puts the fact table in the center and the dimension tables surrounding it, resembling a star (hence the name). A snowflake schema is similar to a star schema but its tables are normalized.

Star schemas are de-normalized data where normalization rules, typical of transactional relational databases, are relaxed during design and implementation. This approach offers simpler and faster queries and access to cube data. However they share the same disadvantage with the non-SQL data bases (discussed below), the rules of data integrity are not strongly enforced.

***Non-SQL Database schema*:** In order to liberate data from relational constraints, several alternate architectures have been devised in the industry as explained in the next sections. These non-traditional architectures include methods that store data in a columnar fashion, or store data in distributed and parallel file systems while others use simple but highly scalable tag-value data structures. The more modern big data storage architectures are known by names like NoSQL,

Hadoop, Cassandra, Lucene, SOLR, Shark and other commercial adaptations of these solutions.

NoSQL database means Not Only SQL. A NoSQL database provides a storage mechanism for data that is modeled in a manner other than the tabular relations constraint of relational databases like SQL Server. The data structures are simple and designed to meet the specific types of data or the analytics use-case, so the data scientist has the choice of selecting the best fit architecture. The database is structured either in a tree, columnar, graph or key-value pair. However, a NoSQL database can support SQL-like queries.

Hadoop is an open-source database framework for storing and processing large data sets on low-cost, commodity hardware. But Hadoop is more than a distributed storage system. It's also a parallel computing platform that is ideal for handling complex data analytics tasks, such as machine learning. Its key components are the Hadoop distributed file systems (HDFS) for storing data over multiple servers and MapReduce for processing the data. Written in Java and developed at Yahoo, Hadoop stores data with redundancy and speeds-up searches over multiple servers. Commercial versions of Hadoop include HortonWorks, MapR and Cloudera. Combined with an open source web UI called, HUE (Hadoop User Experience), it delivers a parallel data storage and computational platform.

Cassandra is another open-source distributed database management system designed to handle large data sets at higher performance. It provides redundancy over distributed server clusters with no single point of failure. Developed at Facebook to power the search function at higher speeds, Cassandra has a hybrid data structure that is a cross between a column-oriented structure and key-value pair. In the key-value pair structure, each row is uniquely identified by a row key. The equivalent of a RDBMS table is stored as rows in Cassandra where each row includes multiple columns. But, unlike a table in an RDBMS, different rows may have different set of columns and a column can be added to a row at any time.

Lucene is an open-source database and data retrieval system that is especially suited for unstructured data or textual information. It allows full text indexing and searching capability on large data sets. It's often used for indexing large volumes of text. Data from many file formats such as .pdfs, HTML, Microsoft Word, and OpenDocument can be indexed by Lucene as long as their textual content can be extracted.

SOLR is a high speed, full-text search platform available as an open-source (Apache Lucene project) program. It is highly scalable offering faceted search and dynamic clustering. SOLR (pronounced "solar") is reportedly the most popular enterprise search engine. It uses Lucene search library as its core and often is used in conjunction with Lucene.

Elastic Search is a powerful, open core search engine based on the Lucene library. It provides a "database" view of data with high performance indexing and search capability. It provides rapid search in diverse data formats, and is in particular suited for textual data searches. In combination with an open visualization tool called Kibana, Elastic Search is ideal for rapid textual data mining, analysis and visualization.

Hive is another open-source Apache project designed as a data warehouse system on top of Hadoop to provide data query, analysis and summarization. Developed initially at Facebook, it's now used by many large content organizations include Netflix and Amazon. It supports a SQL-like query language called HiveQL. A key feature of Hive is indexing to provide accelerated queries, working on compressed data stored in a Hadoop database.

Spark is a modern data analytics platform, a modified version of Hadoop. It's built on the notion that distributed data collections can be cached in memory across multiple cluster nodes for faster processing. Spark fits into the Hadoop distributed file system offering 10 times (for in-disk queries) to 100 times (in-memory queries) faster processing for distributed queries compared to a native Hadoop implementation. It offers tools for queries distributed over in-memory cluster computers that allow applications to run repeated in-memory queries rapidly. Spark is well suited to certain applications such as machine learning (which will be discussed in the next section).

***Real-time vs. Batch Analytics*:** Many of the traditional business intelligence and analytics happen on batch data; a set of data is collected over time and then analytics is performed on data sets that are acquired in batches. In contrast real-time analysis refers to techniques that update information and perform analysis at the same rate as they receive data. This is also referred to as streaming data. Real-time analysis enables timely decision making and control of systems. With real-time analysis, data and results are continuously refreshed and updated. In IoT applications, handling real-time data streaming pipelines are critical to the speed of processing and analysis.

***Data Aggregation & Warehousing*:** Most organizations are replete with disparate databases and data spread all over the firm. Some businesses, have accumulated as many as hundreds, perhaps close to a thousand disjointed Access databases, several SQL servers, data warehouses and diverse file servers stored in various file shares in file servers. Add to that all the files in SharePoint, web portals, internal wiki's and similar content management systems. While implementing data warehouses has been useful to consolidate data, not all of these data files are found in a single data warehouse.

"Data must be liberated, or democratized," as many CIOs define their vision for their data. Business leadership and front-line staff should have access (on an as needed basis) to run analytics across these vast and diverse data storage repositories without having to create a massive data base. As part of data governance activity, the organization must take an inventory of its data warehouses, sources of data and uses of data.

Note the distinction between front-line and back-office processes which create and consume data. Up until now, data was the driver of business transactions. Going forward data analytics will power these workflows.

There is always room to envision data virtualization of all disparate data sources across the enterprise. Data virtualization presents a single and seamless view of all data to you regardless of format and where they're stored.

Ultimately, the goal of data aggregation is to connect to these various databases through a set of plug-and-play connectors. Companies such as Talend, Tableau, Astera, Oracle and Microsoft Data Factory offer such connectors while others like Informatica offer the metadata management functions to create a single virtual view of the entire data assets in your organization.

The ideal analytics systems should be able to "crawl" through an institution's infrastructure and discover data tables and databases residing on different servers. Such a system would locate all Access databases (by file type) or SQL Server databases and allow connection to them by various connectors, for example, ODBC connectors.

1.7 Analytics Layer

In this layer, a data scientist uses a number of engines to implement the analytical functions. Depending on the task at hand, a data scientist may use one or multiple engines to build an analytics application. A more complete layer would include engines for optimization, machine learning, natural language processing, predictive modeling, pattern recognition, classification, inferencing, NLP and semantic analysis.

An *optimization engine* is used to optimize and find the best possible solution to a given problem. An optimization engine is used to identify the best combination of other variables to give an optimal result. Optimization is often used to find lowest cost, the highest utilization or optimal level of care among several possible decision options.

Machine learning is a branch of artificial intelligence that is concerned with construction and building of programs that learn from data. This is the basis for the new wave of AI applications. Machine learning builds adaptive models and algorithms that learn from data as well as adapt their performance to data as data changes over time. For example, models based on machine learning can automatically classify customers, anticipate their needs and make a recommendation. There are three types of machine learning which we'll study in more detail later: unsupervised learning, supervised learning and feedback loop learning. Unsupervised learning trains the model on the historical data to figure out patterns and relationships inherent in the data set. Supervised learning requires labeled data so the model can train and distinguish between different labeled data sets. Feedback learning allows a user to correct the model on the fly, so the model can learn from feedback provided by the user.

Natural language processing (NLP) is a field of computer science and artificial intelligence that builds computer understanding of spoken language and texts. NLP has many applications, but in the context of analyzing unstructured data, an NLP engine can extract relevant structured data from the structured text. When combined with other engines, the extracted text can be analyzed for a variety of applications.

A predictive modeling engine provides the algorithms used for making predictions. This engine would include several statistical and mathematical models that data scientists can use to make predictions. An example is making predictions about patient re-admissions after discharge. Typically, these engines ingest historical data and learn from the data to make predictions. In Part 2, we'll review some of the techniques for predictive modeling in more detail.

Pattern recognition engines, also known as data mining programs, provide tools for data scientists to discover associations and patterns in data. These tools enable data scientists to identify shape and patterns in data, perform correlation analysis and clustering of data in multiple dimensions. Some of these methods identify outliers and anomalies in data which help data scientists identify black-swan[13] events in their data or identify suspicious or unlikely activity and behavior. Using pattern recognition algorithms, data scientists are able to identify inherent associate rules from the data associations. This is called *association rule learning*.

Another technique is building a regression model which works to define a mathematical relationship between data variables that minimizes error. When the data includes discrete numbers, standard regression models work fine. But, when data includes a mix of numbers and categorical data (textual labels), then logistic regression[14] is used. There are linear and non-linear regression models and since many data associations in marketing or biological systems are inherently non-linear, the more complete engines provide non-linear logistic regression methods in addition to linear models.

Classification engines solve the problem of identifying which set of categories a subject or data element belongs. There are two approaches, a supervised method and unsupervised method. The supervised methods use a historical set of data as the training set where prior category membership is known. The unsupervised methods use the inherent associations in the data to define classification categories. The unsupervised classification is also referred to as clustering of data. Classification engines help data scientists to group the key data elements (for example customers, or transactions, and other entities) based on their data attributes.

---

13 A black swan is an unforeseen, unpredictable event, typically one with extreme consequences. The black swan theory is a metaphor that describes an event that comes as a surprise and has a major effect.

14 Logistic regression is a form of regression analysis that includes qualitative values in addition to numerical values. We'll study logistic regression in more detail later.

Inference is the process of reasoning using statistics and artificial intelligence methods to draw a conclusion from data sets. Inference engines include tools and algorithms for data scientists to apply artificial intelligence reasoning methods to their data. Often the result of their inferencing analysis is to answer the question "what should be done next?" where a decision is to be made from observations from data sets. Some inference engines use rule-based approaches that mimic an expert person's process of decision making collected into an expert system. Rules can be applied in a forward chain or backward chain process. In a forward chain process, inference engines start with the known facts and assert new facts until a conclusion is reached. In a backward chain process, inference engines start with a goal and work backward to find the facts needed to be asserted so the goal can be achieved.

Semantic analyzers are analytics engines that build structures and extract concepts from a large set of textual documents. These engines do not require prior semantic understanding of the documents. For example, computer-assisted coding (CAC) applications extract concepts from medical notes and apply semantic analysis to determine polarity and meaning of text; if a diagnosis term is positive, negative, speculative, conditional or hypothetical.

Machine Learning: Another form of data analysis is machine learning. Machine learning is a discipline outgrowth of artificial intelligence. Machine learning methods learn from data patterns and can imitate intelligent decisions, perform complex reasoning and make predictions. Medical predictive analytics use machine learning techniques to make predictions. These algorithms process data at the same rate as they receive data in real time, but they also have a feed-back loop mechanism. The feedback loop takes the output of the analytics system and uses it as input. By processing and comparing their output to the real world outcome, they can fine-tune their learning algorithms and adapt to new changes in data. In this book, we will explore a framework that considers both feed-back loop and feed-forward mechanisms.

Statistical Analysis: Statistical Analysis tools would include descriptive functions such as min, max, mode, median, plus ability to define distribution curve, scatter plot, z-test, percentile calculations, and outlier identification. Additional statistical analysis methods would include regression (trending) analysis, correlation, chi-square, maxima and minima calculations, t-test and F-test, and other methods. For more advanced tools, the R programming language offers a broad range of functions for those serious about statistical testing. One can use an open source tool developed at Stanford called MADLIB, (www.madlib.net). Also, Apache Mahout includes a library of open source data analytics functions. Details of these methods can be found in most statistics text books and is out of scope in this book.

Forecasting and Predictive Analytics: Forecasting and predictive analytics are the new frontiers in data analytics. The simplest approach to forecasting is to apply regression analysis to a calculate regression line and the parameters of the equation line such as the slope and intercept value. Other forecasting methods use interpolation and extrapolation. Advanced forecasting tools offer other types of analyses such as multiple regression, non-linear regression, Analysis of variance (ANOVA) and multi-variable analysis of variance (MANOVA), mean square error (MSE) calculations, and residual calculations. A common technique in medical research is logistic regression. Logistic regression is essentially a regular regression analysis except that the variables in the study can be categorical data.

From a typical regression analysis, the values of slope and intercept can be used as indicators to provide forecasts. A variation of this technique allows identification of outliers as we shall see in the future chapters.

Predictive analytics goes beyond forecasting by providing insight into future events. It is model-driven and includes methods that produce predictions using supervised and unsupervised learning. Some of the methods include neural networks, PCA[15] and Bayesian network algorithms. Predictions require the user to screen the data variables and select the predictive variables and the dependent variables from the prediction screen.

A number of algorithms are available in this category that together provide a rich set of analytics functionality. These algorithms include logistic regression, naive Bayes, decision trees and random forest, regression trees, linear and non-linear regression, time series ARIMA,[16] ARTXp, and Mahout analytics (collaborative filtering, clustering, categorization). Additional advanced statistical analysis tools are often used, such as multivariate logistic regression, Kalman filtering, Association rules, LASSO and Ridge regression, conditional random fields (CRF) methods, and Cox Proportional Hazard models to support text extractions. A brief mathematical overview of these techniques appears in Chapter 5.

One quick definition and clarification about outliers and robustness of advanced analytics models is significant to remember. Robustness means the model is resistant to outliers in data. One of our goals in data science should be finding models that are robust or at least that we understand their sensitivity to outlier data.

Pattern Analysis: Using machine learning and classifier algorithms, researchers can detect patterns in data, perform classification of the subject population, and cluster data by various attributes. Subject population refers to potential consumers, customers or people in any study. The algorithms used in this analysis include various neural networks methods, principal component analysis (PCA), support vector machines, supervised and unsupervised learning methods such as k-means clustering, logistic regression, decision tree, and support vector machines. Patterns give us insights into what data is telling us.

---

15 Principal Component Analysis (PCA) is a technique used to identify the key variables from a data set.

16 Auto Regressive Integrated Moving Average.

Other methods include graphical reasoning, a form of case-based reasoning. These techniques enable a researcher to identify a specific case from the data or identify other cases that match a specific case. For example, this technique can be used to find the right diagnosis for a patient, or match a legal case to a specific legal situation. These methods and their statistical algorithms are explained in more detail in Chapter 5.

1.8 Presentation Layer

This layer includes tools for building dashboards, applications and user-facing applications that display the results of analytics engines. Data scientists often mash up several data visualization widgets, web parts and dashboards (sometimes called Mash boards[17]) on the screen to display the results using infographic reports. These dashboards are active and display data dynamically as the underlying analytics models continuously update the results for dashboards.

Infographic dashboards allow us to visualize data in a more relevant way with better illustrations. These dashboards may combine a variety of charts, graphs and visuals together. Infographic components include different ways of presenting analytics results visually like heat maps, tree maps, bar graphs, a variety of pie charts and parallel charts and many more visualization formats.

Advanced presentation layers include data visualization tools that allows data scientists to easily visualize the results of various analyses (classification, clustering, regression, anomaly detection, etc.) in an interactive and graphical user interface.

Several companies provide rapid data visualization programs including Tableau, QlikView, Panaopticon, Microsoft PowerBI, Amazon Quicksight and Hitachi's Pentaho are revolutionizing how we perform data discovery and visualization. Panopticon has developed a Complex Event Processor (CEP) engine that allows users to view prior events in a graphical representation. 3-Dimensional (3D) and Multi-Dimensional (mD) visualization tools are starting to be adopted in the industry providing rich and rapid view into transactional and operational data.

Open source data visualization components such as D3.js and Kibana are gaining market popularity. For those organizations that have opted to build an all open-source data stack, Hadoop and its eco-system offer many options at each of the four layers described here.

17 Mashboards are defined as a collection of visual graphs that contains different visual components delivered by diverse sets of technologies or analytics engines.

1.9 Data Analytics Process

The data analytics process follows a series of steps, typically along the sequence of the following eight key steps. The first three steps are commonly referred to as Extract-Transfer-Load (ETL). The remaining steps explain the data ingestion into analytics platform and processing the data:

1. Data Extraction. Data extraction is possible via SQL queries or using Apache Sqoop to bring data into a data warehouse or to a data lake (typically Hadoop-based[18]). Data integration with data warehouse systems such as Microsoft SQL Server can be performed via SQL Services Integration Services (SSIS) or SQL Services Report Services (SSRS).
2. Data Transformation. Often data needs to be either normalized or transformed across multiple sources to conform to a common standard, such as a common format or common data elements. Data transformations can occur in Hadoop by data scientists using Hive and the Python programming language.
3. Data Ingestion (Load). In this step the data is properly ingested by the data analytics system and imported into the appropriate data structure. The ingested data can be structured data fields or unstructured text such as physician notes and charting information. Apache Sqoop and Flume are widely used tools to load data into the data lake.
4. Data preparation. Data may be inherently noisy or incomplete. This step normalizes and repairs data preparing it for model building and analysis.
5. Data Discovery (Statistical Analysis). This step helps with understanding the data. Given the frequency and distribution of extracted data, certain statistical inference can be made about the data. Highly correlated data, or unbalanced data must be identified and treated at this step. Such inferences provide information about outliers, trends, clusters of data, correlation analysis and Pareto analysis.
6. Model Building (Pattern Analysis). This step involves building the model, training and testing it against actual data. It may include finding patterns in data by an advanced technique such as artificial neural networks, decision tree, case-based analysis and machine learning algorithms which are used to identify patterns, causal relationships, influence diagrams, classification of data and predictive analysis.

18 Sqoop is typically used with Hadoop ecosystem. Other data storage technologies can be used instead, such as MongoDB, SnowFlake, Denodo, Microsoft Azure Datalake, Amazon S3 and others but the method of data import will vary for each data storage technology.

If the data set is textual and the goal is to develop NLP models, the following steps may be considered:

a. Term Extraction. Using natural language processing (NLP) technologies, the ingested data is parsed and pertinent topic data is extracted from unstructured data. In other use-cases that involve predictive modeling, this step involves building and training the predictive model using quantitative and qualitative values.
b. Semantic Analysis. Once the topic data (terms) are extracted, their relationships, strength and acuity must be established. Semantic analytics provides the intelligent means to identify meaning and relationships between patient and various medical terms.

7. Validation and Meta-Analysis. This stage evaluates the accuracy of the model and steps to improve the validity and quality of results. It might include meta-analysis by considering results of all other analytical steps and models to make an overall set of conclusions and insight about the data that has been analyzed.
8. Visualization. The final step of the process involves presenting the results, typically in graphical dashboards, graphs, and visual charts to tell a compelling story.

Data Mining has received tremendous attention and popularity once its competitive value and early success stories were published. Data Mining attempts to find associations and relationships between data elements that are not obvious. Data analytics uses data mining techniques to identify such relationships and associations between data elements and data sets.

Projects such as Non-Obvious Relationship Awareness (NORA) and data mining software probe databases search for obscure matches and relations between relevant information. This approach can identify connections between data that can span as many as 30 degrees of separation. For example, one degree of separation would be two patients acquiring the same infection in a hospital. The second degree would be finding out that both patients visited the same CT scanner room. A third degree might be discovering that the same transporter transferred those patients. NORA is used to detect fraud and connections between individuals, businesses and activities. It should answer questions about degrees of connections between entities. For example, if I (hypothetically speaking) work at Merck and my wife is an attorney, her law firm may un-knowingly assign her to a legal project at Pfizer, a competitor to Merck, which creates a conflict of interest for both of us. NORA can detect such awkward and conflicting relationships.

Data mining approaches are known as “Black-box” methods where the analyst makes no prior assumptions about the data and inter-relations about the data. The models allow “data to speak for itself.”

There is also a "White-box" approach, where the causal relationships between data are known to the analyst. Data analytics builds upon all these methods and technologies to overcome the inherent complexity of data analysis to extract appropriate and relevant information and then analyze it with both Black-box and White-box methods.

Chapter 2
Basic Data Analysis

2.1 KPIs, Analytics and Business Optimization

The goal of *business optimization* is to use data analytics to navigate through key performance indicators of the organization and identify functions that need further optimization and improvement. At this level, the executive management team can view the business operations through key metrics provided in form of data visualization tools such as dashboards or mashboards. Key metrics can be defined as key performance indicators (KPI). KPIs may include not only direct measurements of performance but metrics that are predictors of the company performance. Predictors can be identified using data analytics methods such as neural networks, signal boosting and correlation analysis. These topics are discussed in the following chapters.

To define KPIs, the analyst selects certain variables, ratios and measurements. These performance indicators might measure certain aspects or business drivers of performance. There are a few KPI templates out there such as quality core measures, or balance score card metrics. Most often these KPIs are tracked, measured using live, real time stream of data and observing the KPI variables over time. Users can set alerts and interact with KPI data in modern dashboards. The user is often able to compose a specific dashboard of KPI variables. The user can drill down on certain variables for more detail or conversely roll-up to a higher level of dashboard display. From the detailed view, the user can select certain variables and perform knowledge discovery and analytics all over again.

A commonly used framework for performance measurement is the balanced score card (an example is shown in Figure 2.1). It brings four aspects of the organization into perspective: financial, operations-process, customer and growth-learning.

Each aspect of the balanced score card consists of specific KPIs. Data analytics techniques such as decision trees and logistic regression can help identify factors that either correlate strongly or predict the KPI values for each of the four areas. Finding these factors can help the organization identify the right KPIs and manage the important variables that contribute to the organization's goals.

Another KPI paradigm called OKRs are in vogue these days since Google's success with this approach.[19] Objectives and key results (OKR) were introduced initially by Intel in the late 1970s. They consist of defining key objectives (what is to be achieved) and key results (how to get to the objective). Each KR is evaluated through a set of measurable results. These measurements provide data that help analyze and track progress toward the objective.

19 *Measure what Matters*. by John Doerr, Penguin Random House, 2018.

https://doi.org/10.1515/9781547401567-003

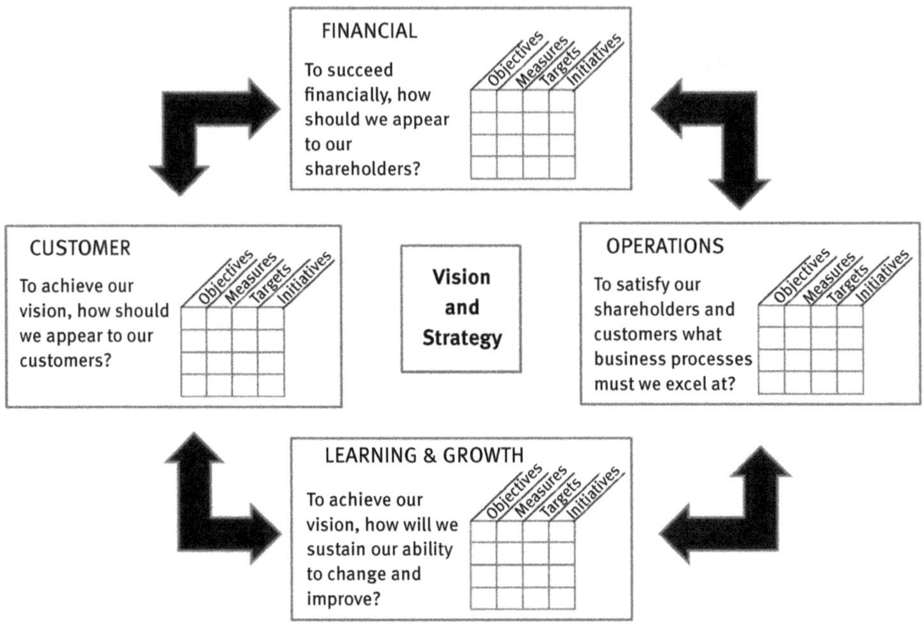

Figure 2.1: Balanced score card as a template for KPIs.

2.2 Key Considerations in Data Analytics Reports

When starting to prepare a data analytics dashboard or report, it's important to begin the data gathering and analysis with the end in mind. Several factors should be considered to determine what data and what type of analysis is required. These factors include:

- **Content:** What will you measure?
- **Audience:** Who is your audience?
- **Production:** Who will be responsible for data and distribution?
- **Accuracy:** How will you ensure the accuracy of information?
- **Frequency:** How frequently should the report be generated? How will it be published and distributed?
- **Format:** What is the ideal format for the dashboard or the report? What aspects of the dashboard should be interactive? What infographic format should be used for visualization?
- **Usage:** How will the information be used?
- **Accountability:** Who is accountable for performance metrics and data reports?
- **Response:** What is the expected and required response to the data presented in the report? What actions are necessary as responses to the report?

2.3 The Four Pillars of a Real World Data Analytics Program

Larger organizations are likely to combine their business intelligence and data analytics plans into a single data analytics program for maximum effectiveness. A typical business intelligence program contains four pillars as described below:

Pillar 1: Data Analytics Strategy

What is your enterprise data strategy? Every data analytics program must define its purpose, goals and objectives and how it intends to support the organization's strategy. Here are some declarative statements that you can craft as the key principles for your organization's data strategy:

1. Data is an asset
2. Data must be shared, be open and accessible but managed, secured and protected
3. Data analytics must give the organization a unique strategic position and a strategic lift
4. Data must integrate our business, not divide it

Data sets must have common definition and dictionary across the enterprise in order to integrate the business lines of business and functional silos. *Strategic lift* means the ability to drastically increase the organization's performance through deep analytics insight. Strategic lift is measured by rise in EPS (Earnings Per Share) and P/E (Price per earnings, also referred to as the stock multiple).

There must be a linkage between the organization's strategy and data analytics strategy. The best practice strategies should answer questions such as:

- How do we monetize data in our organization?
- How can data give our organization a strategic lift?
- How do our data analytics plans map to the organization's strategic goals?
- What is our long range plan for data analytics? The ideal time horizon is to consider a 5-year plan and a 10-year plan.
- What are the strategic plans for analytics? What are the priorities for these analytics objectives?
- How does our analytics plan meet the operational, financial and clinical initiatives?
- How will data analytics be transformative in our organization?

Pillar #2: Data Analytics Governance

Data analytics governance is concerned with establishing policies and scope of data analytics. Best practices recommend forming a data analytics governance

committee as a sub-committee to the IT organization's governance board. This pillar is concerned with answering and guiding the following questions:
- What are our standards for data warehousing and data management across the enterprise?
- What are the security, usage policies and privacy guidelines? Who has ownership to what data and do we have their permission to use that data for analytics?
- What are the proper and appropriate applications of analytics? Specify the improper and banned uses of analytics.
- Who will own the results of analytics and the derivative data and models that are developed from our data? Who will own the intellectual property of the results that we'll discover through analytics?

Pillar #3: Data Analytics Framework

The framework deals with technical aspects of managing data and analytics tools. It answers questions such as:
- What are the infrastructure requirements today and in 5–10 years? Should we build an on premise cloud infrastructure or store data in an off premise private virtual cloud? What are the infrastructure components for data storage and archiving?
- Which systems of record will be supported and designated as analytics platforms? The answer typically starts with existing data warehouse assets and vendor commitments. For example, a business that has invested in Teradata or Oracle and SAS is likely to elect these components as a starting point for its systems of record for analytics. The point is that the source of truth and systems of record must be identified.
- What analytics tools will we support and what will our analytics tools library consist of?
- What technologies and vendor solutions will be supported as enterprise analytics systems to provide infrastructure and analytics tools?
- What is our analytics capability roadmap? What solutions do we intend to deploy in the next five years and in what priority?

Pillar #4: Data Analytics Community

This pillar is concerned with the community of users, meeting their expectations and user's perspective. Questions that the analytics program must address include:
- What are the use-cases and analytics needs plus areas of opportunity for applying analytics?
- Who are the typical users and the user persona for analytics? Personas represent groups of users with similar and common use-cases. Personas may include

management (financial managers, operations manager, sales managers, etc.), knowledge-workers, executives, front-line staff and so on. In a large organization as many as a dozen personas are typical.

- Who produces analytics dashboards (or reports) and who will consume them?
- What are the standards for internal publishing and subscribing to analytics results? The people who produce dashboards and perform the analytics work are typically data scientists and data analysts. In a publish-and-subscribe model, the analysts are publishing their dashboards and the users subscribe to (or consume) those reports.
- What is the expected usage model from the user community? This can define the service level agreement (SLA) between the users and the data analytics group at your organization. The data analytics program should define the role of the user with respect to analytics. Do we provide a self-service model or a DIY (Do-IT-Yourself) model where we provide the infrastructure and analytics tools so the users perform their own analytics and develop their own applications? In contrast, do we provide a full-service concierge desk similar to data marts so that the IT organization performs the analytics and users can consume analytics-as-a-service (AaaS).

Analytics Strategy	Analytics Governance	Analytics Framework	Analytics Community
Strategic Plan Perspective	**Policy Management Perspective**	**Infrastructure Perspective**	**User Community Perspective**
How data analytics plans map to organizational goals?	What are the standards for data warehousing & data management?	What infrastructure components are in scope?	What are the use-cases and who are the users?
What is our long range (5-10year) plan for analytics?	What are the security, privacy, ownership rights & usage policies for analytics?	What are the systems of record for analytics?	Who produces analytics dashboards and who consumes them?
What are the strategic plans for analytics?	What are the guidelines for proper application of analytics?	What will the standard tools library consist of?	What are the standards for internal publishing and subscription of analytic results?
How does analytics meet the operational, financial and clinical initiatives?	Who will own the derivative analytics data and intellectual property?	What technologies will be supported for infrastructure, analytics and visualization?	Usage methods: self-service or concierge?
How will analytics be transformative?		What is our analytics capability roadmap?	Community of analytics users (publishers & subscribers)

Figure 2.2: The four pillars of data analytics program.

Overall the key questions and decisions regarding data management that are needed to be made include five areas:

- What is our data model going to be?
- What is the query model?
- What is our data consistency model?
- What applications and APIs are required to meet our business strategy and analytics needs?
- What is the data analytics user community and how to involve them into the decision making process?

A best practice in data analytics program suggests that we take inventory of data and map them against the organization's analytics needs. A tool for capturing a high level map is using the data analytics matrix. Figure 2.3 illustrates an example data analytics matrix which includes the analytics requirements in the column heading and the rows below it define: the purpose, priority, user community, required data, sources of data, available data warehouses, any data access limitations, data types, data quality, analytics methods required and user interface for

Analytics use case vs. Governance Criteria	Readmission Prediction	Revenue Cycle Leakage Analysis	Clinical Quality Metrics	Population Health Management	Clinical-Translational Medicine Research	Genomics-Precision Medicine Research
Purpose	Predict readmission	Increase revenue capture	Quality reports	Manage population health	Evidence-based medicine	Clinical discovery & advancement
Priority	Med	Hi	Hi	Lo	Med	Hi
User Community	Discharge Nurse	CFO office	Quality & Compliance Office	Primary Care Physician community	Specialists, Clinicians	Life Science researchers
Required Data	Medical billing codes	Medical billing codes + medical records	Medical records	Medical records	Medical records	Genomics data + medical records
Data Source(s)	Billing Systems	Billing systems + EMR	EMR	EMR + Billing systems	EMR	EMR + Genetic sequence data
Data Warehouse	No	No	Yes	Yes	No	Yes
Data Access Issues	ETL possible	ETL possible	Limited	Limited	HL7 extract possible	ETL possible
Data Type (Structured/ Unstructured?)	Structured	Structured	Structured + unstructured	Structured + unstructured	Structured + unstructured	Structured + unstructured
Data Quality	Excellent	Excellent	Good	Average	Average	Good
Analytics methods	Predictive modeling	Rule-based method	Descriptive Statistics	Classification methods	Logistic regression	Bayes Nets, Neural Networks
User interface	Workflow driven dashboards	Audit dashboards	Dashboards with drill-down feature	Dashboards with drill-down feature	Dashboards with drill-down feature	Dashboards with link to medical records

Figure 2.3: Example of a data analytics matrix.

each column. In this example, the illustration shows the formation of a data analytics matrix in the context of clinical and healthcare use-cases.

2.4 The Eight Axioms of Big Data Analytics

So far, we've covered the vast areas of big data, analytics and potential use cases of applying analytics to clinical data. In this final section of Chapter 2, I'd like to share some general observations that I've collected as axioms about big data analytics.

Mathematics has become biology's new microscope and the new telescope for astronomy. The vast amount of data being generated in biology, the physical sciences and business are providing the enormous empirical evidence leading scientists and managers to use mathematical sciences as the new microscope or telescopes for discovery in their field. Just as astronomers in the seventeenth century spent most of their times on building telescopes and less time viewing stars, today's data scientists are spending much of their time curating data leaving less time for analysis. But, that will change as standard analytics platforms will emerge to handle big data volumes. Analytics platforms that provide the standard data "scope" will be widely adopted.

1. **More data is better.** The more data is available, the richer the insight and results that big data analytics can produce. Unlike descriptive statistics which is less interested in outliers and more on the central tendency of data frequency, in big data analytics, outliers and black swan events are of more interest. Our ability to produce knowledge from big data is only limited by our capacity to store and process that data. The bigger and more diverse the data set, the better the analysis can model the real-world. Try to incorporate as many data sets from internal and public sources as possible. There is a corollary to this axiom which is important to note. More data is better, but only to a point. If data is noisy or unbalanced, then too much of it can be a challenge to machine learning and model accuracy. Nevertheless, a data scientist should be able to extract the useful portions of the data from such large datasets if it makes sense.
2. **One model does not fit *all*.** Most scientists and data analysts work hard to find the best model or algorithm that meets their data challenges. But, that approach will fail the challenges for a scalable and extensible that result from the dynamic nature of real world data. As the saying goes, one size does not fit all, a single algorithm does not meet diversity of data sets and changes in data that may occur over time. Hence, as I proposed in my doctoral dissertation, the data scientist must build a committee of models that collectively can produce a more intelligent, adaptable and scalable analytics solution immune to variants of data.
3. **BI and analytics are not the same things.** Where Business intelligence (BI) comes short, big data analytics thrives. When we run business intelligence reports, we're extracting data that are structured, normalized, limited in scope,

often relational and well behaved. The business intelligence techniques fail when we deal with very large and diverse amount of data, unstructured data that's typically found in textual files, noisy or missing data and data that is not in tabular format.

While BI is concerned with descriptive statistics and simple mathematical manipulations of structured data, big data analytics comes with tools such as Natural Language Processing (NLP) to extract data from unstructured text, or with pattern recognition models, predictive models, optimization and machine learning. BI has the power to show snapshots of data presented in graphs. In contrast, big data analytics provides analysis of data through real-time data analysis with feedback that improves the models over time and the results are provided through data visualization tools.

4. **Correlation and causality are not the same.** While positive or negative correlations tell us a lot about relationships between data elements and variables, they don't constitute causality. It takes a lot more to claim one variable is the cause for another. For example, we must show that one variable occurs in time before the other. And, that the spurious variable effects –the indirect effect of other variables- are properly accounted for. But, this is not to discourage seeking the root cause of events, it's to emphasize that interpretive skills are just as important as mathematical modeling talents. Data scientists should have the interpretive skills that go beyond the mathematical abilities to interpret the results of analysis. What do all these correlations tell us and what changes can produce the desired outcomes.
5. **Sparse data analytics approaches will win.** As the volume of data increases, the likelihood that data sets will contain noisy, duplicated data or the opposite, miss important data elements will increase. Hence, whenever we collect massive amounts of data, we should anticipate and prepare to cleanse data, deduplicate and handle sparse data sets. The sparseness will naturally increase as the data volumes rise over time. Therefore, the sparse data analytics methods will eventually overshadow the dense data analytics methods.
6. **Machine learning and data mining are not the same, but cousins.** Machine learning is a branch of artificial intelligence that provides systems that can learn from data. Machine learning is often used to classify data or make predictions, based on known properties in the data learned from historical data that's used for training. Data mining works to provide insights and discovery of unknown properties in the data.

   Machine learning can be carried out through either supervised learning or unsupervised learning methods. The unsupervised learning uses algorithms that operate on unlabeled data, namely, the data input where the desired output is unknown. The goal is to discover structure in the data but not to generalize a mapping between inputs to outputs. The supervised learning use labeled data for training. Labeled data are datasets where the input and outputs are known. The supervised learning method works to generalize a relationship or mapping between inputs to outputs. There is an overlap between the two. Often data mining uses machine learning methods and vice versa, where machine learning can use data mining techniques, such as unsupervised learning to make the engine training more accurate.

7. **Use signal boosting methods.** When data is sparse and the number of data variables grows large, we face a multi-dimensionality challenge.[20] We rely on signal boosting methods to overcome the multi-dimensionality issue to improve model accuracy and performance. Signal boosting increases the significance of certain data variables (which otherwise would have small significance) in data mining and predictive modeling such that their contributions are weighted higher.
8. **Use an ensemble of models.** Studies have shown that combining two or more models can improve accuracy when those models are used in an ensemble. Some models perform better than others on different data sets. For example, some machine learning models can train on small amounts of data better than others and vice versa (some algorithms train better on larger sets of data). How we can create a data analytics algorithm that can work better on any data set? The answer is to build an ensemble (also referred to as a committee of model) so the final result can borrow the best features from each model on a given data set. Even if you average the results of multiple models, you're likely to get a higher accuracy than a single model. This is one of the secrets of the trade that data scientists are starting to notice, and improve their results by using multiple models. For example, when building a predictive model, you can ensemble a random forest algorithm with a neural network algorithm to improve the accuracy of your model prediction.

2.5 Basic Models

The basic models are essentially the OLAP[21] dimensional (cube oriented) type of reports with analysis consisting of slice-and-dice[22] and simple statistics. The user can drill down and roll-up on each slice of multi-dimensional data. These models compute simple but clever management metrics given the user's coding data. Unlike

20 Also known as the dimensionality curse. This challenge occurs when the number of variables grows (often hundreds of dimensions) but the data is sparse making it difficult to make conclusions using statistical significance.

21 Online analytical processing data is structured in a such a way to accelerate dimensional queries of the data.

22 Slice and dice is a systemic approach to break a data set down into smaller parts by reducing the dimensions of data.

traditional graphs that show volumes or quantities, you can include more informative and statistically meaningful annotations in these graphs. These annotations could consist of tick marks that indicate the 50%, 75% or 80% percentiles, one sigma, 2-sigma and 3-sigma variance from the mean, area under the curve, and a slide-bar that allows the user to interact with graphs so that the user can select the percentiles and show results by highlighting a portion of the graph to convey additional information. The basic models consist of time-based reports, quantity-based reports and mixed reports.

These reports may provide additional information inspired by descriptive statistics such min, max, mean, mode, standard deviation, z-score and regression line. Trending and simple predictions can be done by regression line.

2.6 Complexity of Data Analytics

Much of the big data analytics methods are the result of years of research and innovation in artificial intelligence (AI). In artificial intelligence, the difficulty and complexity of solving a problem is defined by its AI-completeness. The most difficult problems are known as AI-complete or AI-hard. This implies that the difficulty of computing these problems is equivalent to solving the central artificial intelligence problem—making computers as intelligent as humans, or strong AI. When a problem is called AI-complete, it implies that the problem cannot be solved by a simple specific algorithm.

According to the *Encyclopedia of Artificial Intelligence,* AI-complete problems include computer vision, natural language processing, problem solving, and knowledge representation, reasoning and dealing with unexpected situations while solving any real world problem. AI-complete problems cannot be solved with modern computers alone. They also require human computation.

Certain machine learning applications combined with a human offer supervised learning that together can solve certain problems related to machine translation or customer sentiment analysis. When the machine learning application makes an error, the user (human computation) corrects the algorithm and the application can "learn" not only from data but also from the user.

If I may oversimplify, AI-easy problems are the class of problems that can be solved in polynomial time.[23] So, how difficult is to solve an AI-complete problem? The answer depends on how complete you want your solution to be. Solving real world problems are still considered AI-complete.

In this book, we'll review algorithms to solve problems ranging from AI-easy to AI-complete.

23 Using a deterministic Turing machine.

2.7 Introduction to Data Analytics Methods

Before we dig deeper into data analytics methods, let's review the three types of data from a statistical perspective: categorical, ordinal, and interval.

Categorical

A categorical variable (sometimes called a nominal variable) is one that has two or more categories, but there is no intrinsic ordering to the categories. For example, gender is a categorical variable having two categories (male and female) and there is no intrinsic ordering to the categories.

Hair color is also a categorical variable having a number of categories (blonde, brown, brunette, red, etc.) and again, there is no agreed way to order these from highest to lowest. A purely categorical variable is one that simply allows you to assign categories but you cannot clearly order the variables. If the variable has a clear ordering, then that variable would be an ordinal variable, as described below.

Ordinal

An ordinal variable is similar to a categorical variable. The difference between the two is that there is a clear ordering of the variables. For example, suppose you have a variable, economic status, with three categories (low, medium and high).

In addition to being able to classify people into these three categories, you can order the categories as low, medium and high. Now consider a variable like educational experience (with values such as elementary school graduate, high school graduate, some college and college graduate). These also can be ordered as elementary school, high school, some college, and college graduate. Even though we can order these from lowest to highest, the spacing between the values may not be the same across the levels of the variables.

Say we assign scores 1, 2, 3 and 4 to these four levels of educational experience and we compare the difference in education between categories one and two with the difference in educational experience between categories two and three, or the difference between categories three and four.

The difference between categories one and two (elementary and high school) is probably much bigger than the difference between categories two and three (high school and some college). In this example, we can order the people in level of educational experience, but the size of the difference between categories is inconsistent (because the spacing between categories one and two is bigger than categories two and three). If these categories were equally spaced, then the variable would be an interval variable.

Interval

An interval variable is similar to an ordinal variable, except that the intervals between the values of the interval variable are equally spaced. For example, suppose you have a variable such as annual income that is measured in dollars, and we have three people who make $10,000, $15,000 and $20,000. The second person makes $5,000 more than the first person and $5,000 less than the third person, and the size of these intervals is the same. If there were two other people who make $90,000 and $95,000, the size of that interval between these two people is also the same ($5,000).

Using machine learning algorithms researchers can detect patterns in data, perform classifications of customer population, and cluster data by various attributes. The algorithms used in this analysis include various neural networks methods, principal component analysis (PCA), Support Vector Machines (SVM[24]), supervised and unsupervised learning methods such as k-means clustering,[25] logistic regression, decision trees, and random forests.

Other methods include graphical reasoning, a form of case-based reasoning. In case-based reasoning, the user is able to search among a repository of prior cases and identify a prior case that matches the current situation. Some use-cases include legal, pharmaceutical, marketing and healthcare.

2.8 Statistical Models

More advanced models include statistical methods that bring new insights about the underlying relationships between the metrics and operational performance of the organization. These models consider relations between variables, correlations and machine learning. We'll study these methods in more detail. The list of models includes regression, correlation analysis, ANOVA, MANOVA, chi squared, and logistic regression. In Chapters 5 and 6, we'll start by covering these statistical methods and then we'll review advanced methods such as classifiers, cluster analysis, data mining, machine learning, prognostics and predictive methods.

---

24 Support Vector Machines is a supervised machine learning technique that separates data into segments by providing a hyperplane that best separates the data segments from each other. It's often used in data classification applications.

25 K-mean clustering is a popular unsupervised machine learning algorithm which segments data into several data clusters by measuring the "distance" between each data point.

2.9 Predictive Analytics

Predictive analytics is intended to provide insight into future events. It includes methods that produce predictions using supervised and unsupervised learning. Some of the methods include neural networks, PCA and Bayesian network[26] algorithms. Predictions require the user to select the predictive variables and the dependent variables from the prediction screen.

A number of algorithms are available in this category that provide a rich set of analytics functionality. These algorithms which are explained in the next paragraphs include logistic regression, naive Bayes, decision trees and random forest, regression trees, linear and non-linear regression, time series ARIMA,[27] ARTXp, and Mahout analytics (collaborative filtering, clustering, and categorization).

Additional advanced statistical analysis tools are often used, such as multivariate logistic regression, Kalman filtering, association rules, LASSO and ridge regression, conditional random fields (CRF) methods, and Cox proportional hazard models. Add to this list many tools for concept extraction, topic modeling and text mining that support natural language processing and understanding. A brief mathematical overview of these techniques appears in the next section.

2.10 Advanced Analytics Methods

The following section is a peek into a common list of analytics algorithms used by data scientists.

- ***Linear Regression*** – The goal of linear regression is to find a linear equation that best fits a set of data. A data point may consist of multiple dependent variables. For example, when there are two dependent variables, the best fit equation is a plane through the observed dependent data values. For example, a user might be interested in computing the regression graph to predict a customer's volume of purchase using three independent variables, customer's age, education level and income level. A regression curve can provide an estimate of what a prospect is likely to spend.
- ***Kalman Filtering*** – This technique is used to predict the future value of certain dependent variables over time. It works by using the most recent values of a variable and adjust the prediction by minimizing error between the predicted and observed result. The result is fed back into the model to make the next prediction at a lower error value. For example, it can be used to predict short term

26 Bayesian networks are probabilistic graphical diagrams based on the Bayesian theorem that represent a set of variables and their conditional probabilities in a directed acyclic graph.
27 Auto Regressive Integrated Moving Average.

values of a patient's blood pressure, temperature and other physiological measurements if adequate historical data are accumulated. It's also a good technique to handle noisy and dirty data as we'll see later.

- ***LASSO*** – A regression analysis method that combines both variable selection and regularization in order to enhance the accuracy of prediction. It's a regularized version of least squares method (The least absolute shrinkage and selection operator). LASSO is preferred for computing compressed sensing, a technique in signal processing that acquires and reconstructs a signal by solving the undetermined linear systems. Thus the entire signal can be reconstructed from a relatively few measurements. It's applications in healthcare include improving MRI and CT image processing.

  LASSO methods are used in pattern recognition in images and improve predictive analytics. One of the applications of LASSO is variable selection in those situations when you have a large collection of covariates and you want to select a set for the efficient prediction of a response variable. (A covariate is a variable that is possibly predictive of the outcome under study. It can be a direct predictor or a confounding or interacting variable that affects the predicted outcome.)
- ***Multivariate logistic regression*** – The multivariate logistic regression is a form of regression analysis used for predicting the result of a categorical dependent variable subject to one or more predictor variables. An example of categorical is gender (MALE, FEMALE). This example results in a binomial or binary logistic regression since the outcome can have only two possible values. Other categories may have been created from continuous data. For example, weight can be categorized as LOW, MED and HIGH based on certain thresholds. Logistic regression is used extensively in healthcare and medical research. One of the widely used methods to predict mortality in injured patients called trauma and injury severity score (TRISS) is an example.[28]
- ***Association rules*** – A technique used in machine learning to develop association rules between texts when data mining. Association rule mining helps find patterns in data by identifying features (data dimensions) that occur together and features (data dimensions) which are correlated. It helps find what the value of one feature implies about the value of another feature. It's often used in market basket analysis. For example, if a shopper purchases chips, what's the likelihood that the shopper also purchases salsa?
- ***Ridge regression*** – Also referred to as Tikhonov regularization, is a form of least squares method. It has been used in medical research to predict patient

28 Boyd, C. R., Tolson, M. A., Copes, W. S., "Evaluating Trauma Care: The TRISS Method. Trauma Score and the Injury Severity Score," *J Trauma*. Vol. 27, No. 4, 370–378, Apr. 1987. URL: http://www.ncbi.nlm.nih.gov/pubmed/3106646

diseases and their stage of disease growth. For example, it has been used to predict which stage of disease is a patient will be in the next few months or predict cancer recurrence in patients.

- ***Decision trees and random forest*** – Decision tree and its cousin influence diagrams are similar graphical methods for calculating expected values (or expected utility) of competing alternative decisions. Random forests are essentially ensembles of decision trees. They have many applications in classification and prediction. We'll review these techniques in more detail in the next chapter.
- ***Time series ARIMA*** – Autoregressive-integrated-moving-average models are used to predict the next value in a given time series dataset. Given a time series data set, the ARIMA model uses two parts, an autoregressive (AR) component and a moving average (MA) component to predict the future values in the series. They have many applications to detect a sudden change of frequency or rate. For example, in cyber security, time series analysis is used to determine threat activity from normal activity. In the healthcare industry, they are used to detecting abnormally high visit rates related to a particular disease that may be an early signal of an outbreak, epidemic or bioterrorist attack.
- ***ARTXp*** – The autoregressive tree with cross connect model is similar to ARIMA except that it uses a decision tree algorithm to predict the next value in periodic time series data. If the user is interested in finding only the next value in a recurring time series data set, ARTXp is recommended.
- ***CRF methods*** – Conditional random fields are excellent for pattern recognition and machine learning since they take context into account. Often used for parsing sequential data, they can predict the next word in a sequence of phrases. For example, in medical research, CRF methods which are considered among supervised machine learning techniques have been used to train models to identify negation successfully.
- ***Cox proportional hazard models*** – Proportional hazard models are a form of survival model. These models calculate the time that passes before some event occurs to one or more covariates that are associated with the outcome.
- ***Naïve Bayes method*** – The goal of naïve Bayes (NB) method is to classify data using the simple probabilistic Bayes' theorem with strong (naïve) independence assumptions. This method is popular for classification and diagnostics. For example, applications for diagnosis and prediction in healthcare use it because it is simple with good accuracy and can explain how it arrived at a classification. NB methods have been used in medical research to predict and diagnose a number of diseases ranging from cancers to hepatitis and liver disorders.
- ***ANOVA/MANOVA*** – Analysis of variance and multivariate analysis of variance methods measure how congruent or incongruent the variance changes in variables with respect to each other. These tools are often used to determine the impact of a product or marketing campaign on a population segment in comparison to a control group.

- ***Principal component analysis*** – The PCA method is closely related to factor analysis. It is excellent for classification and prediction of independent variables and their contribution to the outcome (dependent variable). It's a method that translates correlated variables into uncorrelated variables. PCA is used in medical research to identify the key principle factors that affect or classify a patient's disease. It is used in genetics to identify the internal structure of a given genetic data set and diseases, as well as identify which datasets are similar.

  In other industries, the PCA method is used to predict when equipment is likely to fail. For example, in airline industry, data from aircraft engines are analyzed to predict when a preventative repair is needed.

We'll review advanced analytics and natural language processing models in more detail in Chapters 4–11. But, first let's review data analytics processes in the next chapter.

Chapter 3
Data Analytics Process

3.1 A Survey of Data Analytics Process

Data analytics is the process of decomposing data as a whole into separate components for individual examination and/or combining the components into the whole for the purpose of generating insights.[29]

Leading data analytics vendors and academic institutions have proposed several process models in the last 20 years. Companies such as SAS, IBM, Oracle and Microsoft have offered process models that focus on certain aspects of the analytics work, mostly from a business perspective.

Oracle and Cloudera have informally offered a 7-step "value-chain" process to drive value from the data analytics process.[30] This process tends to highlight a value-chain methodology and offers some valid points. These steps include:

1. **Decide on the objectives:** As the first step in the value-chain, the business unit must decide on objectives for data science teams. This step contemplates the success measures and key performance indicators early in the process.
2. **Identify business levers:** This step identifies the key metrics and decisions that can improve business performance. It's a step to connect business goals to metrics early in the project.
3. **Data collection:** This step involves casting a wide net to collect data from diverse sources, both internal and external.
4. **Data cleaning:** Almost all analytics processes recognize this step as necessary to improve data quality, handle missing data and remove noise from data.
5. **Data modeling:** In this step data scientists build models designed to correlate data with business outcomes to identify insights and make recommendations.
6. **Grow a data science team:** This step emphasizes building strong teams: data science teams to focus on modeling and predictions plus data engineering teams to focus on building data ingestion and data curation pipelines.
7. **Optimize and repeat:** The data value chain is perceived as a repeatable process leading to continuous improvement. Based on the results of the model, the

29 Another definition for data analytics: "Data science is a multi-disciplinary field that combines skills in software engineering and statistics with domain experience to support the end-to-end analysis of large and diverse data sets, ultimately uncovering value for an organization and then communicating it to stakeholders as actionable results."

30 "The Seven Key Steps of Data Analysis," Gwen Shapira, *Oracle PROFIT newsletter*, published by Oracle Corp, May 23, 2013.

https://doi.org/10.1515/9781547401567-004

business applies changes to the business levers and measures the impact of those changes. The faster the business can repeat this cycle, the sooner it can get value from its data.

Another data science process model developed by Dr. Carol Anne Hargreaves suggests a 7-step business analytics process.[31] This process model shown in Figure 3.1 provides a business-centric perspective on the analytics activity. An advantage of this process is its focus on defining the business need first and then applying the data analysis methods. This process ensures that the results of analysis are actionable and the "decision" is applied to update the system.

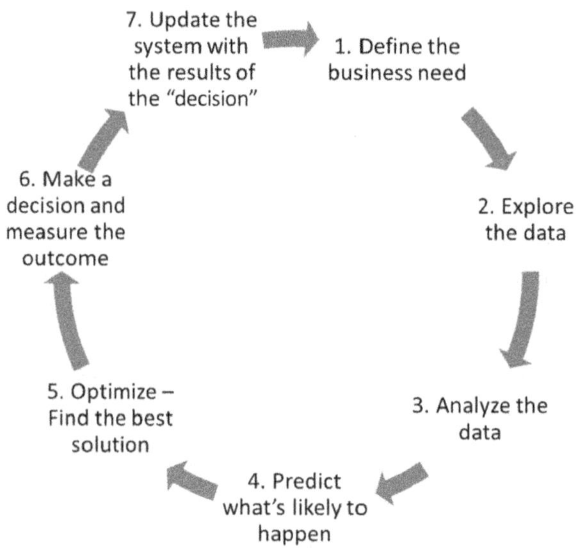

Figure 3.1: The 7-step data analytics life cycle process model.

While this process model offers a good outline of activities, it assumes a limited range of analytical practices, mostly confined to the hypothesis testing and prediction phases. A wide range of other data analytics methods, machine learning and textual analytics to name a few, are missing.

The steps include:

1. **Define the business need:** This stage involves understanding the business goals and objectives or the problem to be solved. It identifies business stakeholders, what data is available and what is the deliverable from the analysis.

31 "The 7-Step Business Analytics Process," Dr. Carol Anne Hargreaves, *NUS-ISS Quarterly e-newsletter*, No. 3 (Jul–Sep 2013). Institute of Systems Science, National University of Singapore.

2. **Explore the data:** This stage includes data cleansing, making imputations for missing data, refining and normalizing data as necessary. It involves transforming data to create new metrics or variables.
3. **Analyze the data:** In this stage, the data scientist applies statistical methods and algorithms to identify patterns and correlations among data variables. It could involve hypothesis testing, correlation analysis or regression.
4. **Predict what's likely to happen:** This step emphasizes techniques to uncover insights and relationships that can help predict an outcome.
5. **Optimize-find the best solution:** In this step the data scientist runs "what-if" scenarios and algorithms to find the optimal solution. Accuracy improvement belongs to this step.
6. **Make a decision and measure the outcome:** The goal of this step is to put the insights derived from analysis into action and measure the outcome.
7. **Update the system with the results of the "decision":** This step compares the outcome measures of the decision against the expected outcomes. It determines the impact and efficacy of the data analysis.

Another process model, is offered by Cathy O'Neil and Rachel Schutt. This model, illustrated below, starts with collection of raw data from the world, followed by

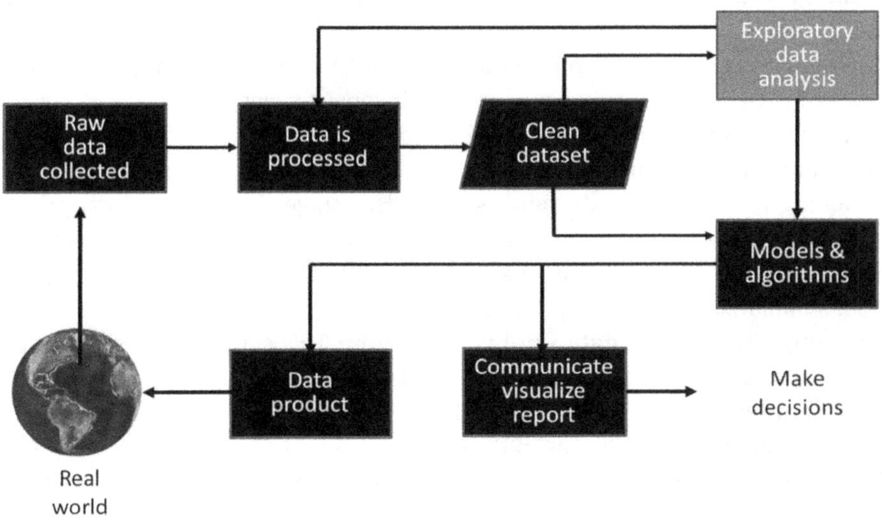

Figure 3.2: Data science process model.

processing and cleansing the data.[32] It considers two possible scenarios for analysis: explanatory data analysis and model development. The results can be communicated via visual reports or as a data product that automates the entire process delivering the result in a product.

There are three other data analytics process frameworks that merit mention: CRISP-DM, SEMMA and KDD. While these process models were developed a long time ago and with data mining practice in mind (not data science practice), they're worth a brief review to establish a baseline. We'll review these process models next. A good review of all three process models are given by Umair Shafique and Haseeb Qaiser.[33]

3.2 KDD—Knowledge Discovery Databases

The Knowledge Discovery in Databases (KDD) process shown in Figure 3.3 was designed to find knowledge in data with an emphasis on high level data mining methods. It consists of 9 steps.

1. **Developing and understanding of the application domain:** In the first stage of the KDD process, the project goals are defined from customer's perspective and used to understand the application domain and existing knowledge.
2. **Creating a target data set:** This stage focuses on creating the target dataset, data samples and variables.
3. **Data cleaning and pre-processing:** This stage involves cleaning and pre-processing the data to remove any noise or inconsistencies in data.
4. **Data transformation:** This stage focuses on transforming data and its format to prepare data for analysis in such a way that can be used by algorithms.
5. **Choosing the suitable data mining task:** This stage chooses the appropriate data mining task based on the goals defined in the first stage. Tasks are defined as categories or methods of analysis, such as: classification, clustering, regression, summarization, optimization, etc.
6. **Choosing the suitable data mining algorithm:** This stage selects the appropriate data mining algorithms to meet the task in stage 5.
7. **Employing data mining algorithm:** This stage applies the selected algorithm that best identifies the patterns in the data.
8. **Interpreting mined patterns:** This stage focuses on interpreting and evaluating the mined patterns. The practitioner may present a visualization of the results in this stage.

32 "Doing Data Science," by Cathy O'Neil and Rachel Schutt, 2013.

33 "A Comparative Study of Data Mining Process Models (KDD, CRISP-DM, and SEMMA)," Umair Shafique and Haseeb Qaiser, Dept. of Information Technology, University of Gujrat, *International Journal of Innovation and Scientific Research*, ISSR Journal, Vol. 12, No. 1, 217–222, Nov. 2014.

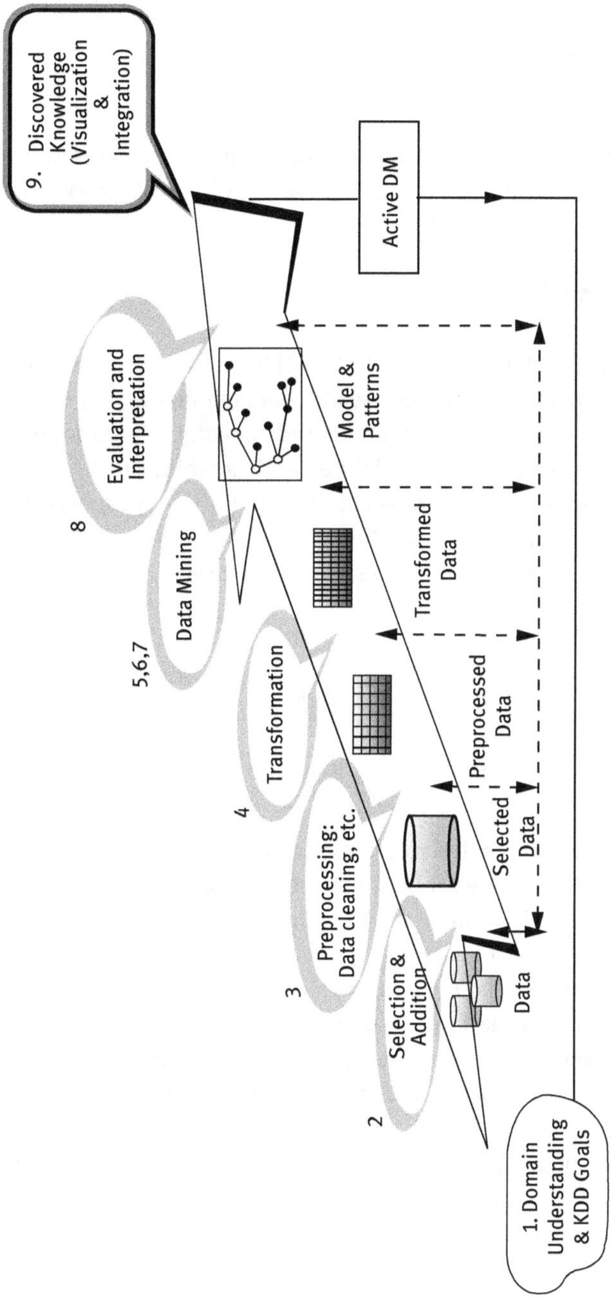

Figure 3.3: KDD data analytics process.

9. **Using discovered knowledge:** In the final stage of the process, the discovered knowledge is used for various purposes, and prepared to be integrated with other systems or further action.

The KDD process model has been predominately used for data mining use-cases in the past. It was the predecessor to and in some extent influenced the CRISP-DM model.

3.3 CRISP-DM Process Model

Cross Industry Standard Process for Data Mining (CRISP-DM) is an open standard process model that describes the common approaches used for data mining.

The purpose of CRISP-DM is to provide a reliable and repeatable data mining process. It offers a uniform framework with guidelines and flexibility to adapt to different business problems and different data sets.[34] It was initially proposed by Daimler-Benz, SPSS (now IBM) and NCR back in 1996.

CRISP-DM breaks the data mining process into six major phases:

1. Business understanding
2. Data understanding
3. Data preparation
4. Modeling
5. Evaluation
6. Deployment

These six steps (shown in Figure 3.4) are described in more detail below:

1. **Business understanding:** This phase is about understanding project objectives and requirements. It recommends: statement of business objective, data mining objective, and success criteria. It focuses on understanding the project objectives, requirements and what the client wants to accomplish.

   This step defines the current situation, the business problem, expected benefits of the solution. The practitioner determines the resources (hardware & software) available, the data mining goals, an estimate of effort, the critical steps and criteria for model assessment.
2. **Data understanding:** This phase prescribes acquiring the data, exploring the data and verifying the quality of data. It focuses on understanding the general properties, volume and type of data. Data collection occurs in this phase.

34 CRISP-DM 1.0 – Step-by-Step Data Mining Guide, Pete Chapman, Julian Clinton et. al. https://www.scribd.com/document/264461662/CRISP-DM-1-0-Step-By-Step-Data-Mining-Guide. Also see "The CRISP-DM Model: The New Blueprinting for DataMining," Colin Shearer, *Journal of Data Warehousing*, Vol. 5, No. 4, 13–22, 2000.

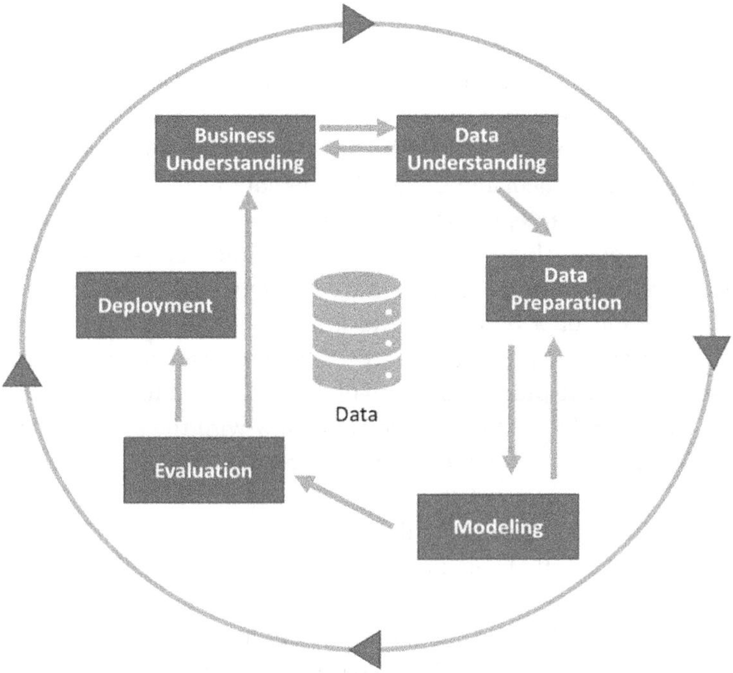

Figure 3.4: CRISP-DM data analytics process model.

This step includes data exploration: identifying the data attributes, range, correlation and attribute type. For each attribute, basic statistics tests (distribution, max, min, mean and standard deviation, variance, mode and skewness) are performed.

The practitioner identifies special data values, errors, missing attributes, plausibility of values and data relationships in this phase.

3. **Data preparation:** This phase prepares data for analysis. Activities in this phase are generally data transformations. They include cleaning the data, integrating and joining the data, normalizing data, rearranging attributes, reformatting data and reordering records.
4. **Modeling:** This phase involves selecting and building the model, generating tests and evaluating the effectiveness of the model.

   The practitioner evaluates various modeling techniques and selects the right model based on its fit for the data parameters and attributes. Next, the model is applied to the data. Some post-processing of the results may be necessary. Finally, the reliability and plausibility of results are analyzed.
5. **Evaluation:** This phase evaluates the model more thoroughly and determines how to use the results. The model evaluation considers error rates, robustness of the model, interpretation of results, any business issues overlooked or introduced by results and the final decision based on the model results.

In this step, results are ranked, any failures and issues with the model are resolved. The practitioner evaluates the results to understand them and their impact against the objectives set in phase I. Finally, determine if the decision can be made to proceed to the next stage.

6. **Deployment:** This phase determines how the results need to be utilized, who will use them, how to deploy them and the frequency that they'll be used.

   The practitioner develops a deployment plan, including an ongoing maintenance and monitoring plan. This phase deploys the results and sends entire package into production. The results may be utilized as business rules, an interactive on-line application or a database. The knowledge gained from these phases are to be organized and presented in a way that the customer can use it.

   As the final step, the entire process is documented into a final report and the project post-mortem is conducted (a final project review with the stakeholders to determine lessons learned and what could have been done better).

In CRISP-DM, phases 4–6 may be repeated as data or business deployment options change over time. While some non-IBM data mining professionals use CRISP-DM, IBM remains the primary company that embraces this process model. Generally, while CRISP-DM is still being used and IBM has announced an extension to it, this process model seems to be outdated, not meeting the practical needs of the modern data scientists.

3.4 The SEMMA Process Model

The SEMMA process model stands for Sample, Explore, Modify, Model and Access. It was a data mining method presented by the SAS Institute. It enables data organization, discovery, development and maintenance of data mining projects. Its five stages offer the following:

1. Sample: This is the first but optional stage of SEMMA. It involves sampling of data to extract critical information about the data while small enough to manipulate quickly.
2. Explore: This stage focuses on data exploration and discovery. It improves understanding of the data, finding trends and anomalies in the data.
3. Modify: This stage involves modification of data through data transformations, and the model selection process. This stage may evaluate anomalies, outliers and variable reduction in data.
4. Model: The goal of this stage is to build the model and apply the model to the data. Different modeling techniques may be applied as datasets have different attributes.
5. Access: The final stage of SEMMA evaluates the reliability and usefulness of the results, performance and accuracy of the model(s).

In general, the three data mining process models have a lot in common. Some of the phases in one process correspond to similar phases in another. For example, the KDD process stage "data transformation" can correspond to the "modify" stage of SEMMA process and to the "data preparation" stage in CRISP-DM.

While these process models are still being practiced in a few organizations, their useful shelf-life is almost over if not completely. The industry demands a modern and appropriate data analytics process standard that reflects the work of the new generation of data scientists. We propose DSMAT as the new process standard.

3.5 Microsoft TDSP Framework

Microsoft announced a Team Data Science Process (TDSP) in 2017 as an agile, iterative data science methodology to deliver advanced analytics solutions and intelligent applications efficiently.[35]

The goal of TDSP is to improve collaboration, learning and best practices to successfully facilitate data science initiatives. The main component of TDSP is the data science lifecycle.

TDSP data science lifecycle definition: The lifecycle perspective supports a product- oriented approach for data scientists who ship their solution as part of intelligent applications. The lifecycle offers rigor and repeatable steps through five major stages (see Figure 3.5). Each stage must address goals, how to carry out the step, and the next steps.

The five steps of Microsoft's TDSP life cycle include:

1. **Business understanding:** The goal of this step is to define business objectives, stakeholders and the business problems. It finds relevant data that helps answer the business questions. The key objective of this step is to identify the key business variables and the analysis needed to meet the purpose of the project.
2. **Data acquisition and understanding:** The goal of this step is to produce a clean, high-quality data set relevant to the target variables. It develops a data pipeline to regularly refresh data.
3. **Modeling:** The modeling stage determines the optimal data features for the algorithm, model building, model training and validation to be deployed into production.
4. **Deployment:** This stage operationalizes the model. Once the model is ready for deployment, it's deployed as a product with pipeline to production.
5. **Customer acceptance:** The goal of this step is to confirm that the data pipeline, the model and their deployment into production meet customer's acceptance

35 "What is the Team Data Science Process?" Microsoft Azure documentation, October 19, 2017. Contributors: Mark Tab, Gary Ericson, Josee Martens, Craig Casey, Sheri Gilley.

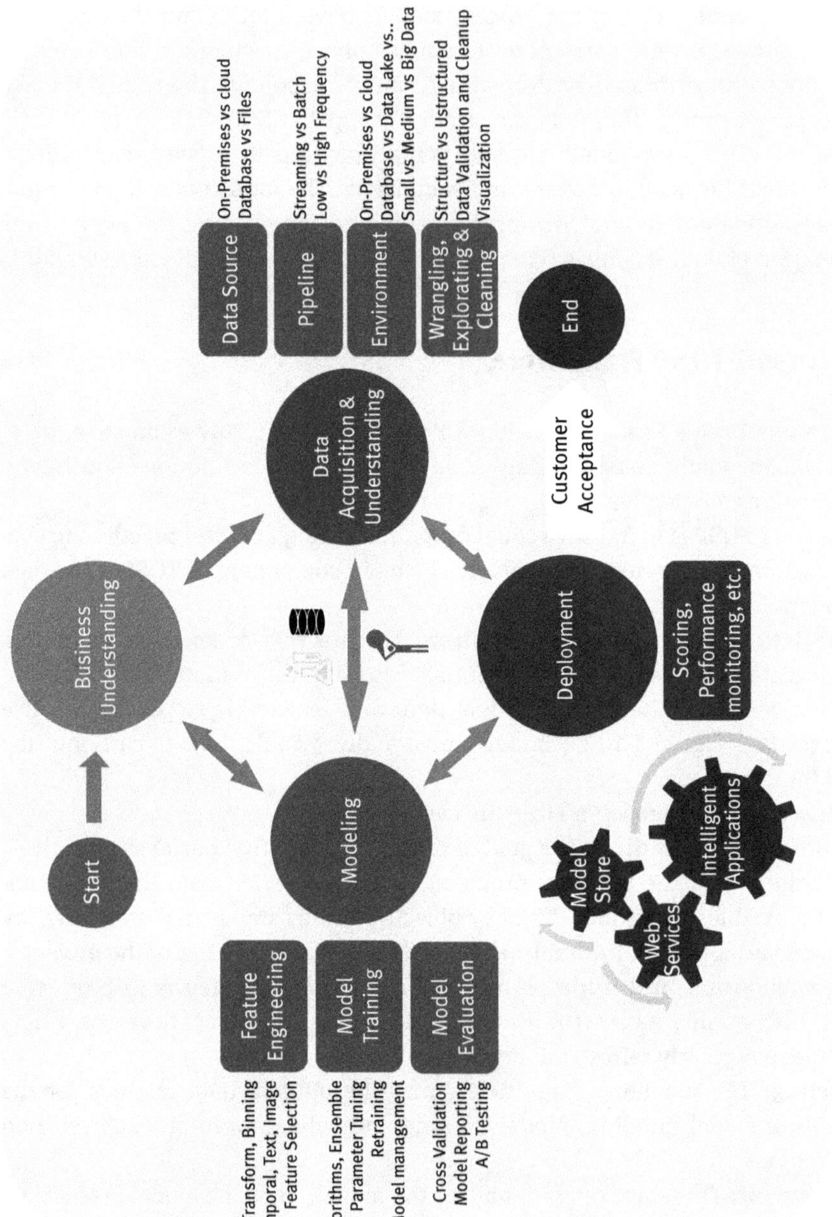

Figure 3.5: Data Science Lifecycle – Microsoft TDSP Model.

and objectives. It includes a validation activity and project hand-off to the entity who will run the system in production.

The TDSP offers a lifecycle approach which provides a sustainable and repeatable process. Our goal in this project started with developing a process framework that defines what a data scientist is expected to do and help data scientists be more successful in their work. Although most of these process frameworks address some gaps or needs in the industry, they've not gained the industry-wide acceptance as standards.

The following is an illustration of a common data science process for a predictive modeling project.

3.6 Data Analytics Process Example—Predictive Modeling Case Study

In this section, we'll illustrate a typical data science process through a marketing analytics example. Our scenario predicts consumer purchase likelihood and we'll take the journey of analysis from start to the end. The scenario that we'll illustrate is looking to predict consumer purchase of goods. For example, consider the scenario where you want to predict which customers are likely to buy new golf clubs in response to an advertisement. The assumption here is that the data scientist is using R and in particular the DMwR[36] package for data prep and analysis.

Consumer Purchase Prediction—Marketing Analytics Scenario

Marketing organizations are interested to target consumers for their marketing campaigns and predict which consumers are likely to select a product. The goal of this project is to identify the consumers who will make the purchase. The result of our analysis will indicate the probability of purchase for each applicant. We'll apply data analytics methods and processes to this problem starting with curating data. The following is a summary of the methodology applied to this project.

Understanding the Problem

Predicting the probability of purchase is critical to any marketing campaign. The data comes from a marketing survey. Consumers have responded to a marketing survey that measures the consumer's interest in golf. For example, the frequency

36 Data Mining with R, https://cran.r-project.org/web/packages/DMwR/

that they play, the frequency that they go to tournaments, watch golf tournament events, whether they take golf lessons and membership to gold magazines, their favorite golfer, etc. Using this data, our goal is to determine if the consumer is likely to purchase a certain brand of golf equipment. Each question on the survey can provide an important clue, a signal that can be used in analysis.

Data Curation

As is often common in the data science practice, the data set available from surveys has many inconsistencies and issues. There are missing values, outliers and noisy data fields. These issues must be handled before using the data for the model. The data set consists of 10,000 records and 30 attributes, features that we'll use for machine learning exercise next.

Feature Discovery and Selection

Of the 30 attributes, it's likely that a few applicant parameters make more significant contribute to the prediction of purchase. As a result, those features must be discovered (feature discovery) and selected for the final model training.

Model Development

There are many analytical models available including deep learning (using neural networks), logistic regression, principal component analysis or using random forest (an ensemble of decision trees). We'll use a decision tree for this model and evaluate the performance (accuracy of results) to validate our choice.

Accuracy Evaluation

We'll evaluate the model performance (quality, consistency, and robustness) using various tools such as AUROC (explained in more detail in future chapters). If the accuracy is not acceptable, we'll adopt another model.

Overall Validation

We'll analyze results from multiple perspectives including validity and generalizability, ethical issues and balanced perspective before releasing the model into production.

Data Prep Steps

The data set has some missing data fields that must be imputed. We'll discuss data imputation in detail later. We'll impute data by generating new values for missing data fields from the other complete records available. But, first we normalize numerical data. Then, we use the k-nearest neighbor algorithm to perform multiple imputation. The k-nearest neighbor finds the most plausible value for a missing data item by comparing all other data fields across all records. The data scientist uses the knnImputation () function of the DMwR package for this purpose.

Considering that the data set has about 10,000 records, there can be outliers in the data set. As part of data discovery and prep, we're interested in further analysis of data and possibly removal of outliers. The data scientist may use qualitative features to detect outliers using the Levels() function. To detect numeric data outliers, we normalize the data between [0,1] and then we apply the Daisy() function of the cluster package. The function uses a boxplot technique to identify outliers. Next we need to rank the outliers. This is accomplished by techniques such as the agglomerative hierarchical clustering algorithm available by using outliers.ranking() function in DMwR.

After ranking the outlier data, the data scientist may choose to remove outlier records that are out of range or fill the outlier values with NULL.

Data inconsistencies are a big barrier to analysis. Data may be unbalanced. For example, if 70% of survey respondents are male and the remainder are female, the data is unbalanced and the machine learning algorithm is likely to train better on the male population than on the female population. The data scientist applies tools such as the Smote() function to balance data. We also randomly select a portion of the dataset (around 80%) for training the model and 20% of the data for testing. The balancing function must be applied to the training data segment.

Feature Discovery

Before we begin model-building, the first step is to identify any data relationships that are problematic such as highly correlated data fields. Highly correlated data fields introduce redundant information represented by two or more attributes. The data scientist can use the PlotCorr() function to identify and remove highly correlated data fields. This function is available in the Ellipse package.

This step helps with discovery of important features (predictive variables) and ranking the features. The data scientist ranks the attributes (features) based on some threshold limits, the highly ranked features will be selected for model building. In order to rank the features, the data scientist may use tools such as the RandomForest() function available in the RandomForest package. This package includes a function called RFCV() which help with ranking the features.

Model Building

The new reduced dataset with a reduced feature set is now ready to be put into a model. Data classification technique is one approach for prediction. Essentially, the data scientist builds an algorithm to classify the dataset into two groups (those who will purchase and those who won't purchase). The model is trained on the training data set to learn from key features (predictors) to predict how a consumer should be classified. Decision trees are commonly used algorithms for classification. We're using labeled data for training. This means that we know which respondents made a purchase in the past. As an alternative to Decision trees, the data scientist could have chosen other algorithms, such as a binary logistic regression approach, or neural network, or support vector machines. The data scientist chose the decision trees approach based on historical success for data classification of this nature.

The data scientist then trained the model on the training data set (80% of records randomly selected) to build the decision trees model. For this step, the rpart() function can be used. Next, the model is applied to the test data set (20% of records) to determine the accuracy and performance of the model. To test the model, the predict() function may be used.

Performance Evaluation

The results of the test are compared against the actual, original classification labels to determine the accuracy of the model. By comparing the results with the original test data labels, the data scientist determines the number of true positives, true negatives, false positives and false negatives. Accuracy evaluation may consider the precision, true positive rate, true negative rate, and simple accuracy ratios. Ultimately, additional model adjustments or other models may be used to compare accuracy improvement. We'll review accuracy measures and how to improve model performance in the upcoming chapters in more detail.

Part II: **Advanced Analytics Methods**

Chapter 4
Natural Language Processing

4.1 Natural Language Processing (NLP)

The idea of natural language processing (NLP) is to develop a computer system that can analyze, understand and generate natural human language. There are many applications in NLP including topic modeling, customer sentiment analysis, machine translation and concept search.

There are four categories of NLP techniques:

1) Pattern matching: This approach attempts to interpret input text by patterns rather than combining the structure and meaning of words.
2) Syntactically driven parsing: This approach studies how words can fit together to form higher level units such as phrases, clauses and sentences.
3) Semantic grammars: The semantic approach parses the language into semantic categories that are defined by their meaning.
4) Case frame instantiation: This parsing technique combines bottom-up recognition of key language components with top-down instantiation of less structured components.

A central task of NLP is parsing. Parsing is the process of computing valid structures of a string of text, or a list of tokens according to a given grammar. Since a sentence can be written in many valid structures, parsing a sentence to extract meaning is not simple. It's not surprising that natural language processing is regarded as an AI-complete problem.

Processing NLP has many challenges. The ultimate goal is to train the model to convert a piece of text into a computer friendly format and for humans to have "conversations" with computers. We want to train the model to understand and learn natural language.

The NLP tasks include the following, in the order of going from raw text to classified text for understanding:

- Part of speech tagging (POS): Identifies syntactic roles of individual words into their syntax identities such as noun, adverb, verb, etc.
- Chunking (CHUNK): Defines the syntactic constituents by taking chunks of words, such as noun phrase, verb phrase, etc.
- Name entity recognition (NER): Identifies entities such as people, company, location, etc. by tagging parts of sentences by the type of entity.
- Semantic role labeling (SRL): Identifies the semantic role of the words and tags them with the appropriate role. This is the hardest task. The goal is to find all possible verbs and label the arguments of the relation (the subject and object) to the verb.

https://doi.org/10.1515/9781547401567-005

Computer programs are trained to identify the correct tags and labels for the parsed text. An analysis of Wall Street Journal articles identified about one million distinct labeled words through POS, chunking and NER tasks. Reuters found about 200K labeled words using NER. The accuracy of some models is reported ranging from 88% to 97% for these corpuses of text. Given that the inter-human understanding error is believed to be around 15%, these levels of accuracy are remarkable.

The problem of natural language processing consists of three stages and with each stage there is versatility and several choices of tools and methods. These stages are:

1. ***Parsing***
2. ***Learning***
3. ***Semantic Modeling***

Parsing choices – Parsing methods include grammars and inference procedures. The approaches include inductive logic programming (ILP), synchronous context free grammars (SCFG), combinatory categorical grammars (CCG) and probabilistic context-free grammar. Simply put, context free grammars consist of a rule for substituting one word for another. For example, a rule to replace "A" with "α" in a word like "Apple" will produce "αpple." Another rule might reverse the string, such that "Apple" becomes "elppA." Probabilistic CCG measures the frequency of derivations (or permutations) that are possible from a given string. Of these methods, I'll only cover CCG and PCFG methods.

Learning methods – The choice of learning approach is to decide on which supervised learning technique we want to apply. There are several techniques such as annotated parse trees introduced in 1994 and sentence-LF pairs. More recent approaches use latent variable methods including question-answer pairs, instruction-demonstration pairs, conversation logs and visual sensors. There are both supervised and semi-supervised learning approaches. I'll cover the neural networks and learning approaches in more detail later in this book.

Semantic Modeling – Here we're concerned with what logical language representation to use and how we go about modeling the meaning of the text. There are a number of options for semantic model as well such as *variable free logic* and *high-order logic*. The more recent methods include relational algebra and graphical models.

There are several tools available to handle NLP. NLTK is the most popular Python text mining toolkit. There are other options including OpenNLP and TextBlob.

NLTK includes an easy user interface and over 50 corpora and lexical resources such as WordNet and several text processing libraries for classification, tokenization, stemming, tagging, parsing and semantic reasoning. NLTK includes *NLTK Data* which is a library of corpora, grammars, models and tools for tagging. An example of Corpora is the *Brown Corpus*. NLTK includes a tool for sentence boundary detection and sentence segmentation called *Sent_Tokenize*. Similar tools for word chunking and tagging are called *Word_Tokenize* and *POS_Tag*.

Tokenizers are used to divide sentences into lists of substrings and substrings into list of words. For example, Sent-Tokenize is used to find the list of sentences and Word-Tokenize to find the list of words in strings.

Tokenizing sentences is often known as *sentence boundary disambiguation* (SBD), sentence boundary detection or sentence segmentation. Sentence tokenization determines where sentences begin and end.

Again there are many tools for sentence tokenization such as NLTK, OpenNLP, TextBlob, MBSP, SpaCy and such. If you're using Python language, NLTK is the tool of choice. NLTK includes pre-trained tokenizers for over 17 languages. You can find the list of supported languages on the NLTK website (www.nltk.org).

Many word tokenizers (including NLTK's word tokenizer) use a standard tokenizer called TreeBankWordTokenizer. The alternative word tokenizers are the PunktWordTokenizer and WordPunktTokenizer. The big difference between tokenizers is in how they treat punctuation. PunktTokenizer splits the sentence on punctuation but keeps it with the word, while PunktWordTokenizer splits all punctuations into separate tokens. For example, consider the sentence: "This's a test". The TreebankWordTokenizer, splits the sentence into: ['This', "'s", 'a', 'test']. Using PunkWordTokenizer, the same sentence is tokenized into ['This', "'", 's', 'a', 'test'].

Part of Speech (POS) tagging is an important task in natural language processing. POS is also called *grammatical tagging* or *word-category disambiguation*. It's the process of labeling a word according to its definition and context in relation to other words in the sentence. For example, POS tagging identifies the subject, object, verb and prepositions in a sentence automatically.

Text processing includes stemming and lemmatization. Stemming is the process of reducing inflected and derived words to their stem, base or root form. For example, "walking", "walks" and "walked" have the stem "walk". The stem need not be identical to the morphological root of the word. Usually, a related word may map to the same stem even though the stem itself is not a valid root. Some search engines treat words with the same stem as synonyms. This process, called *conflation* is used as a kind of query expansion, namely extends the search to include the search term as well as the synonyms for the search term.

Lemmatization is the process of grouping together the different inflected forms of a word so they can be analyzed as a single item. It's the process of finding the lemma for a given word. Lemmatization requires an understanding of the context of the word and determining the part of speech for a word in a sentence. Lemmatization requires knowledge of the grammar of a language.

Lemmatization is closely related to stemming but the difference is that a stemmer operates on a single word without knowledge of context (or grammar), so it cannot discriminate between words that have different meanings. Thus stemmers are easier to implement. Lemmatization, finds the root of the word based on its semantic meaning (context) and part of speech in the sentence.

NLTK offers several stemmer tools, such as Porter stemmer, Lancaster stemmer, and Snowball stemmer which is based on the Snowball stemming algorithm. You can import these stemmers in order to stem words in a text.

NLTK uses WordNet for lemmatization. WordNet is a large lexical database of English. It groups nouns, verbs, adjectives and adverbs into set of cognitive synonyms (called synsets). These groupings express distinct concept. Synsets are interlinked by conceptual semantics and lexical relations. The result is a network of meaningfully related words and concepts that can be searched or navigated with a browser.

WordNet has several advantages. It may resemble a thesaurus, but it's more than that. First it groups and links words together to make sense of the words. As a result, words that appear in close proximity to one another in a network are semantically disambiguated. Second, WordNet labels the semantic relations among words in these grouping. In contrast, a thesaurus only groups words based on their meaning similarity not based on any explicit pattern in text.

Named Entity Recognition (NER) is the process of detecting persons, companies, and locations from text. Some of the useful tools for NER are included in the Stanford NER with Python. You may also use Stanford POS Tagger and Stanford Parser in NLTK since NLTK now provides interfaces to all three tools. One version of Stanford NER recognizes four classes of data (Person, Organization, Location, misc.). A 7-class model recognizes up to seven classes of entities including: Time, Location, Organization, Person, Money, Percent and Date.

Text classification is another important application of NLP. The goal of text classification is to assign a label to one or more categories of documents. It's useful in library science, information science and even sentiment analysis. NLTK can be used to classify text into categories of documents. Text classification is enhanced through training the model, also known as supervised learning. NLTK offers many algorithms for supervised learning such as naïve Bayes, decision trees and maximum entropy models (also known as multinominal logistic regression). I'll cover these models in future chapters.

As we mentioned earlier one of the popular parsing grammars used for parsing, formalizing and semantic understanding of natural languages include combinatory categorical grammar (CCG) and probabilistic context-free grammar (PCFG). We'll study CCG and PCFG parsing in the coming sections. For a complete end-to-end set of tools, you may visit NLP sites for Cornel and University of Washington Semantic Parsing Framework (SPF).[37]

37 http://yoavartzi.com/spf

4.2 NLP Capability Maturity Model

As the natural language processing applications become more complex, they require increasing proficiency in advanced capabilities. I've depicted an NLP capability maturity model in the next diagram (in a ladder format). It illustrates how the increasing capability in NLP processing enables more complex and powerful applications.

The model begins with tokenizing, breaking sentences into words. The applications are simply searching and word counting. Word counting is an important technique in quantitative analysis.

The next level is adding POS and tagging. In this step, the data scientist adds language grammars and syntactical tags to words. The applications at this level are top modeling and sentiment analysis. Topic modeling enables us to define the type of word as Person, Company, Location and even the type of words.[38] Sentiment analysis informs us about the direction and sentiment of the sentence. The direction of the sentence is known as polarity. Polarity of the sentence determines if the sentence is positive, neutral, negative, etc.

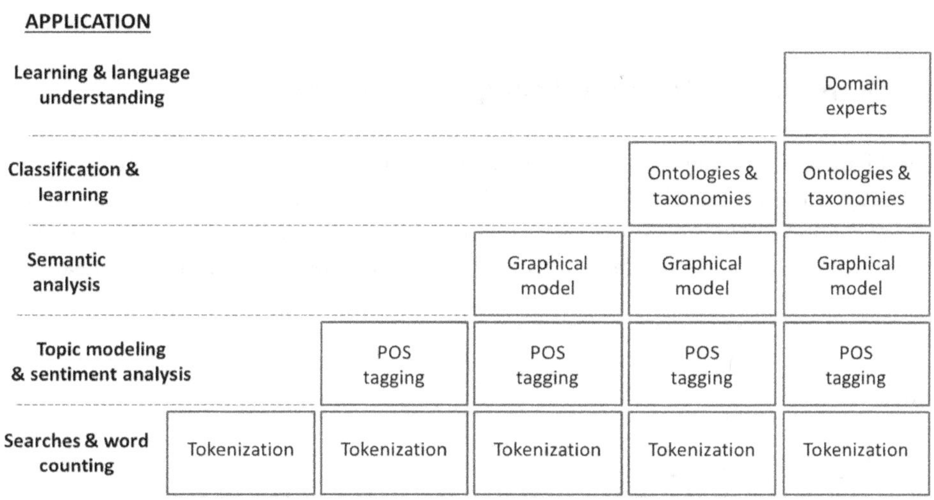

Figure 4.1: NLP capability maturity model.

The next step in capability maturity model brings additional techniques, such as graphical models and case-based inference. These techniques attempt to deliver semantic meaning from text.

38 To define types of word means the ability to determine if a word is a location, or name of a person or an organization, or a product, etc.

Going beyond semantic analysis requires a combination of analytical capabilities and methods including machine learning and classification. In order to enhance NLP results, we need to introduce layers of ontology and taxonomy. Taxonomy is a library of words specific to a certain usage and application. For example, a library of words, acronyms and synonyms widely used at your firm constitutes a taxonomy. Ontology is a broader, industry-wide library of words that are used across an industry. Both allow better understanding of context in a text.

The highest layer of capability is natural language understanding. This layer relies on all underlying capabilities. It works to understand the language, its content and context. Developing domain experts is a technique to enhance understanding. Domain experts are a collection of expert knowledge frames and rules that mimic how an individual processes text or spoken words. The ability to process natural language is a tacit knowledge. Domain experts capture the tacit knowledge into a set of rules and frames that provide the context for understanding. For example, if the intent is to understand medical notes, we need to develop capabilities along medical ontologies (such as UMLS, LOINC and ICD-10 libraries), plus the medical domain expert capability which defines rules about how the text should be interpreted.

4.3 Introduction to Natural Language Processing

The history of NLP begins with work developed by Franky Rosenblatt in 1957 who introduced the notion of a perceptron. The perceptron received a number of inputs (X_i) and applied some weights (W_i), to compute an output Y, as shown in the diagram below. However, this model has a huge disadvantage. It was simply linear.

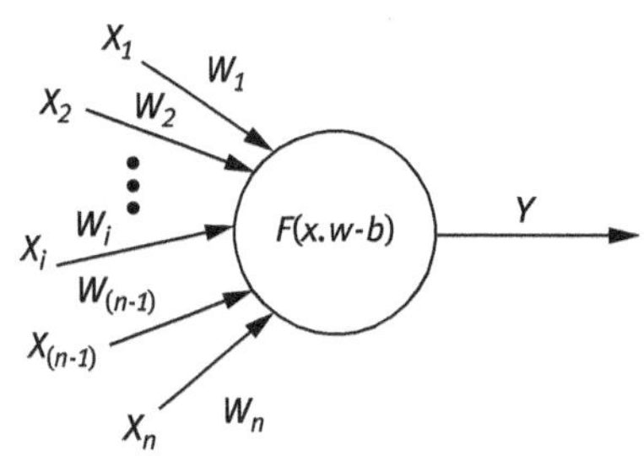

Figure 4.2: A simple perceptron.

There were many attempts to make the model non-linear. But it was until 1986 when artificial neural networks (ANNs) were introduced by Rumelhart, Hinton and Williams. The neural networks, which I'll cover in more detail in the following chapters, consist of multi-layer perceptrons. ANNs provided multiple layers of computation: an input layer, hidden layer and output layer. They also introduced several methods such as back-propagation and feed-forward to adjust the weights, thereby simply training the model.

Later, other techniques such as support vector machines (SVMs) were introduced but they generally carry a high computational cost. Neural nets are elegant models for machine learning and ideal for constructing what is now called *deep learning*. Furthermore, variants of neural nets called *Convolution Neural Networks* (CNN) are becoming popular in natural language processing.

Much of big data collected in this decade is either machine data or human generated text in unstructured format. More than 80% of the data is in form of narrative text and notes, not in concrete data fields. It would be a major advantage to have a natural language processing (NLP) tool integrated into the analytics process. The ideal NLP tool not only extracts data but also presents the extraction along with relationships to other data to create an interim structured dataset which can be used for further processing.

The ideal NLP tool is equipped with some semantic processing capability. Simple semantic processing functions would include synonyms, identify associations as defined by the user, exclusions (such as exclude all negative cases or certain phrases) and provide a simple frequency count of phrase occurrence.

Advanced semantic processing will support meaning extraction by clustering phrases by building graphs of sentence structures. Semantic processors map these structures to concepts in domain-specific ontology models. The user will be able to define the variants of a noun using synonyms or phrases that most frequently appear with the noun.

For example to define variants of the phrase "ocular complications," the user (or the tool might suggest) entering synonyms and variants such as: ocular, oculars, oculus, oculi, eyepiece, ophthalmic, ophthalmia, optic, optics, complication, complications, and so on.

As another example, when a user searches for "Use of thermogram in detection of meningitis," the semantic feature of the tool would identify meningitis as a disease and thermogram as a method. Notice a representative semantic sentence structure built by a semantic analyzer that might look like a structure in Figure 4.3. These internal structures maintain relationships and meaning that are critical to intelligent term extraction from medical records.

Two important tools that enhance context and natural language understanding is lexicon and taxonomy. A lexicon is a collection of terms used in a particular domain. Lexicons help us related the tokenized text to specific domain of knowledge. For example, the lexicon (collection of words or dictionary) in the legal domain is unique and different from a lexicon in pharmaceutical industry. Taxonomy includes specific words but also includes classifications and the rules associated with each

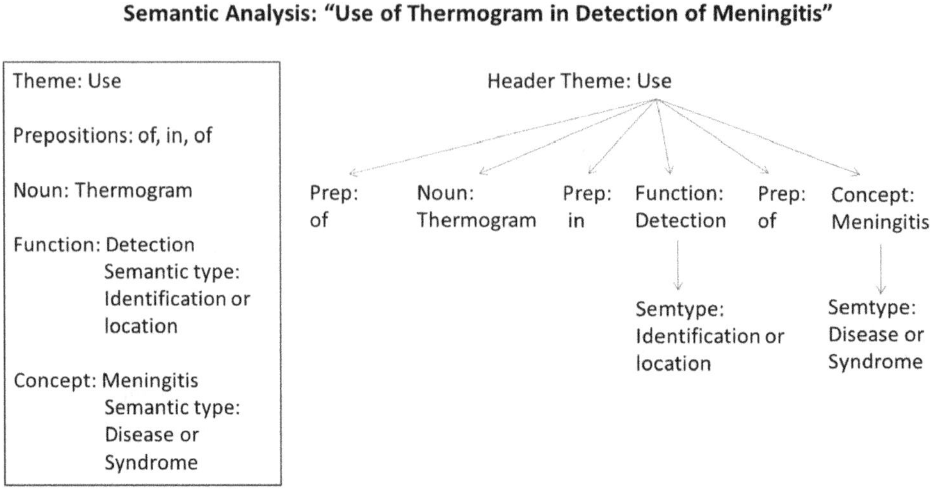

Figure 4.3: Example of possible semantic analysis of a search phrase.

word. In medical applications, taxonomies such as UMLS[39] and MeSH[40] are commonly used by data scientists to build NLP applications.

4.4 NLP Techniques—Topic Modeling

Topic modeling is the science of extracting topics from textual information. Topic modeling is a form of text mining. Its purpose is to identify patterns in a corpus. In other words, it's a method for finding and tracking clusters of works (known as "topics") in large bodies of text.

Just as you go through an article with a highlighter and highlight phrases or passages of text, topic modeling attempts to identify key words of themes. Each set of words is a topic.

There are many applications of topic modeling that answer questions such as: What is the text about? What is the summary of the text? What is the sentiment hidden in the text?

A basic and widely used method in topic modeling is latent Dirichlet allocation (LDA) which compares the occurrence of topics within a document. It's used to identify the probability that a term (or a topic) appears in a document. An open source tool called MALLET uses LDA method as its core algorithm.

39 Unified Medical Language System is a compendium of biomedical terms.

40 Medical Subject Heading (MeSH) is a set of medical terms and classifications of terms.

In order to make the topic modeling more effective, experts recommend that you compile a large corpus of documents, around 1,000 pieces of text. Next you should tokenize the text by transforming it from sentences to strings of words, strip out the punctuation, stop words (such as "the", "a", "and", "of", etc.).

The topic models need to be trained on your data. You can specify how many topics the model should identify. Selecting the number of topics to be extracted and that matters whether you specify 5, 50 or 500 topics. You can run the model multiple times, test and tune the model through trial and error until the desired result is achieved, namely adequate topics are identified with acceptable accuracy.

Topic modeling allows us to "zoom in" and "zoom out" of the document by identify themes within it. We can narrow the scope of themes to broaden the themes. We can track to see how themes have changed over time. For example, by topic modeling newspaper articles, blog posts, twitter and other social media, we can determine public opinion, customer sentiments and how people's attitudes towards certain subjects are changing towards that topic.

Probabilistic topic modeling is a suite of algorithms that work to discover and annotate large corpus of documents with thematic information. These are statistical methods that analyze the words in the texts to discover the themes and how the themes connect to each other and how they change over time.

The idea behind the LDA algorithm is that documents contain multiple topics. A topic is defined as a distribution over a fixed vocabulary. We generate the words in a two-stage process:

1) Randomly choose a distribution (namely a frequency of occurrence) of topics
2) For each word in the document:
    a) Randomly choose a topic from the distribution over topics in step #1
    b) Randomly choose a word from the corresponding distribution over the vocabulary

The statistical nature of topic modeling works as follows: Each document exhibits topics in different proportion (step #1); each word in each document is drawn from one of the topics (step #2b); where the selected topic is chosen from the per-document distribution over topics (step #2a).

4.5 NLP—Names Entity Recognition (NER)

Named entity recognition (NER) is the procedure to identify Persons, Locations, Organizations, Products[41] and entities of interest. It's one of the easiest statistical

41 Persons refers to the proper nouns or names of individuals. For example, George Clooney or Humphrey Bogart. Location refers to a geographic entity, such as a river like the Nile or a city like

linguistic methods. For example consider the statement below as input to the NER program.

INPUT:
"Profits soared at Boeing Co., easily topping forecasts on Wall Street, as their CEO Alan Mulally announced first quarter results."

The NER Tagging program will tag the text with specific entity types. For example, consider the following output from the NER program.

OUTPUT:
Profits/NA soared/NA at/NA Boeing/SC Co./CC ,/NA easily/NA topping/NA forecasts/NA on/NA Wall/SL Street/CL ,/NA as/NA their/NA CEO/NA Alan/SP Mulally/CP announced/NA first/NA quarter/NA results/NA ./NA

Where the tagging schema used by the NER program includes the following tags:
NA = No Entity
SC = Start Company
CC = Continue Company
SL = Start Location
CL = Continue Location
SP = Start Person
CP = Continue Person

The result of this study determines the company, person and locations identified in the text.

RESULT:
Company: Boeing
Person: Alan Mulally
Location: Wall Street

4.6 NLP—Part of Speech (POS) Tagging

Another NLP technique is *part of speech* (POS) tagging which can provide more information. The output in POS tagging will look like the following.

Paris. Organizations are typically names of companies, like IBM or International Business Machines. Products refer to products that carry a brand recognition such as Pepsi or Aspirin. A NER is able to detect these proper names from the text.

OUTPUT:
Profits/N soared/V at/P Boeing/N Co./N,/, easily/ADV topping/V forecasts/N on/P Wall/N Street/N,/, as/P their/POSS CEO/N Alan/N Mulally/N announced/V first/ADJ quarter/N results/N ./.

Let's apply POS tagging using syntactic structure. Given the input below, we construct a syntactic structure.

INPUT:
Boeing is located in Seattle.
Where the tagging used include:
N = Noun
V = Verb
P = Preposition
Adv = Adverb
Adj = Adjective
POSS = Possessive

The tree de-composition of the text using tags is shown in Figure 4.4.

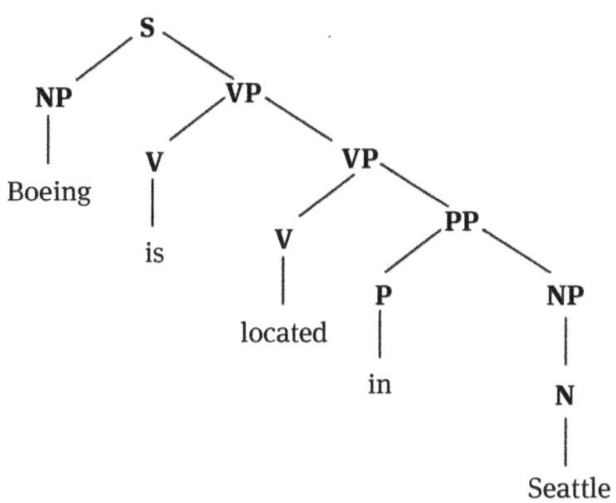

Figure 4.4: POS tagging tree.

Another application of NLP is machine translation. For example, the following text is input and the German translation appears as output.

INPUT (English):
Boeing is located in Seattle. Alan Mulally is the CEO.

OUTPUT (German):
Boeing ist in Seattle. Alan Mulally ist der CEO.

The Penn Treebank Project provides phrase tagging nomenclature for English. For Chinese, the Penn Chinese Treebank is used. For German the NEGRA corpus is used for tagging phrases in text.

Some of the useful online links are: http://www.nltk.org/book/ch05.html, http://nlp.stanford.edu/ and http://www.cis.upenn.edu/~treebank/.
The English tagging taxonomy includes the following:

1. CC Coordinating conjunction
2. CD Cardinal number
3. DT Determiner
4. EX Existential there
5. FW Foreign word
6. IN Preposition or subordinating conjunction
7. JJ Adjective
8. JJR Adjective, comparative
9. JJS Adjective, superlative
10. LS List item marker
11. MD Modal
12. NN Noun, singular or mass
13. NNS Noun, plural
14. NNP Proper noun, singular
15. NNPS Proper noun, plural
16. PDT Pre-determiner
17. POS Possessive ending
18. PRP Personal pronoun
19. PRP$ Possessive pronoun
20. RB Adverb
21. RBR Adverb, comparative
22. RBS Adverb, superlative
23. RP Particle
24. SYM Symbol
25. TO to
26. UH Interjection
27. VB Verb, base form
28. VBD Verb, past tense
29. VBG Verb, gerund or present participle
30. VBN Verb, past participle
31. VBP Verb, non3rd person singular present
32. VBZ Verb, 3rd person singular present
33. WDT Wh-determiner (Wh indicates interrogative word, or question word since they often start with “wh”)
34. WP Wh-pronoun

35. WP$ Possessive wh-pronoun
36. WRB Wh-adverb

4.7 NLP—Probabilistic Context-Free Grammars (PCFG)

Probabilistic approaches consider the probability of words occurring in text in various forms. The probabilistic context free grammars (PCFG) algorithms discussed earlier, work as follows. Suppose you're given a sentence S. We want to find the most likely parse τ. How are we supposed to find $P(\tau)$?

Essentially, we're looking for probability of term τ over the probability of Sentence S in the text. The computation follows this equation:

$$\begin{aligned}\arg\max_{\tau} P(\tau|S) &= \arg\max_{\tau} \frac{P(\tau,S)}{P(S)} \\ &= \arg\max_{\tau} P(\tau,S) \\ &= \arg\max_{\tau} P(\tau) \quad \text{If } S=\text{yield}(\tau)\end{aligned}$$

The challenge is that there are infinitely many trees in the language! We want to compute the probability of τ namely $P(\tau)$. So, we define probability distributions over the rules in the grammar, completely context free (hence the name of the method).

Consider the example below (Figure 4.5) that shows the probability of certain tags occurring in Sentence S.

S	→	NP VP	0.8
S	→	S conj S	0.2
NP	→	Noun	0.2
NP	→	Det Noun	0.4
NP	→	NP PP	0.2
NP	→	NP conj NP	0.2
VP	→	Verb	0.4
VP	→	Verb NP	0.3
VP	→	Verb NP NP	0.1
VP	→	VP PP	0.2
PP	→	P NP	1.0

Figure 4.5: Probability measure of tags in sentence.

In PCFG analysis, it's important to note that the probability of the tree is the product of the probability of the rules that created it. Consider a Sentence S for example: Mary drinks coffee with cream. The PCFG graph of this sentence looks like the graph shown in Figure 4.6.

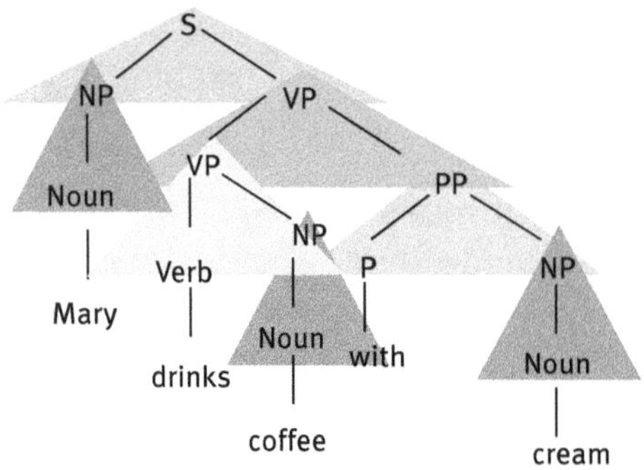

Figure 4.6: PCFG graph example.

The context-free probability for the sentence S is computed as follows:

$$P(\tau) = 0.8 \times 0.3 \times 0.2 \times 1.0 \times 0.2^3 = 0.00384$$

The probabilities for all other sentence structures are similarly computed. There is more research coming in natural language processing. Among the new research that promises new capabilities are:

- Coarse-to-*fine* parsing: This technique uses parsing with a simpler grammar. Then it refines the parsing with a more complex grammar
- Dependency parsing: It works by relating each word to another in a sentence. In dependency parsing, a sentence is parsed by relating each word to other words in the sentence which depend on it.
  - Discriminative parsing: This approach attempts to classify sentences by learning from some training examples. Given a set of training examples, discriminative parsing learns a function that classifies a sentence with its parse tree.
  - Generative models: This approach works to simultaneously extract syntactic structure and name entity relationships at the same time. This technique has shown more than 80% accuracy in finding relationships.

4.8 NLP Learning Method

The purpose of learning in NLP processing is to feed some text into the model and output the appropriate tagging (be it CCG or another parsing method). An approach to learning is word embedding and using neural networks to train the model to find the most appropriate tag for each word.

Supervised learning uses pre-existing labeled text as a reference to train the model. The semi-supervised and "weak-supervision" methods require less labeling expertise. The labels don't uniquely determine correct logical forms. The logical forms in weak supervision are latent. Hence, learning requires executing logical forms within a system and evaluating the result.

As we'll study later, there are different neural network models that can be used for learning including convolution neural networks. These models will be discussed in the deep learning chapters.

4.9 Word Embedding and Neural Networks

There are two approaches to NLP. One is hand-crafting the word relationships and their semantic relationships (as we learned through CCG and PCFG). The other is to use "large scale" approaches that typically apply machine learning and graph analysis.

In a large-scale machine learning approach, we start understanding NLP from scratch (without prior knowledge about text). The goal of a large-scale machine learning technique is to find unified hidden representations of text and their relationships.

We begin a large-scale approach by word embedding. Word embedding[42] is the process of embedding words in a vector space. We use machine learning to train on embeddings. Once the words are embedded, we use an input window to identify the correct tag for each word. *Windowing* is the process of sampling a few words together; say three words at a time. The window shifts through the text one word at a time with the goal of guessing what the correct tag for the middle word should be. For long sentences, where the verbs might be further outside of the window, it's important to consider longer slider windows.

The alternative to longer window size is to feed the entire sentence into a convolution neural network. Said differently, we feed the entire sentence into the network and we run convolutions to handle the variable-length of each sentence. Each convolution produces a probability of a tag for each word. Then we take the maximum values over time to capture the most relevant tags as features. This method produces local word features with higher levels of abstraction.

4.10 Semantic Modeling Using Graph Analysis Technique

One of the techniques to analyze text and build semantic models is to form a relation graph between words that appear in different sentences and paragraphs. A graph can simply be defined as nodes (representing words) and edges (representing

42 The idea behind word embedding is to capture as much of semantic, hierarchical and morphological information about words as possible. The process of word embedding is to map a set of words into a vector of numbers. For simplicity, consider a vector that contains a "1" when a word is present and a "0" when that word is absent in a sentence. This is the basis for techniques such as Google's Word2Vec functions.

relationships) that connect nodes. Some of the recent publications in this area are published on research from GraphLab project and Google.[43]

The graph technique works by parsing text into words and turning each word into its stem. For example, we change the plural of a noun to a singular form and verbs into their roots. Words by themselves are atomic. We want to study them in pairs (bi-grams) or in windows of 3 words or more. An n-gram takes words in collection of one to *n* words, such as a noun, adjective and verb. Generally, an n-gram is a continuous sequence of n items from a given sequence of text.

The windowing technique is the process of shifting the window of n-words through the text and connecting words that have the same root in the document. The result is a graph that connects the words together across sentences and paragraphs in the document.

The resulting graph depicts how words in text appear relative to each other and how the words share their meaning. Each graph consists of a node (represented by the word) and an edge (represented by a relationship). A node that has more inbound links implies having more importance and meaning in the document. That node (or word) gains higher importance and as a result higher ranking in the document. This is known as *word ranking*.

One interesting property of graphs is that they can be converted to matrix format where we can apply matrix algebra calculations. A matrix can be formed by labeling the rows and columns based on the words in the text and place a '1' or '0' if there is a relationship between the two notes (as shown in Figure 4.7).

Higher ranking words can represent abstractions of what the text is about, namely the concept summary of the text. Another related technique is to identify nouns that appear immediately before and after each word and build a graph using those nouns.

This technique is known to deliver a better semantic representation of the text. Figure 4.8 illustrates an example of tokenized text and a graph of the most recurring nouns. The words adjacent to high ranked words are underscored to show their relationship with each other. The example illustrates that the concepts extracted are about Honda, Quality, Problem, Supplier and Recall.

We should note two other tools that are important to text mining and NLP analysis: TextBlob and Pattern. TextBlob is a Python library for NLP processing. It includes

43 *PowerGraph: Distributed Graph-Parallel Computation in Natural Graphs*, by J. Gonzalez, Y. Low, H. Gu, D. Bickson, C. Guestrin (graphlab.org/files/osdi2012-gonzalez-low-gu-bickson-guestrin.pdf

Pregel: Large-scale Graph Computing at Google, by Grzegorz Czajkowski et. al. (googleresearch.blogspot.com/2009/06/large-scale-graph-computing-at-google.html)

Topic modeling with LDA: MLib meets GraphX, by Joseph Bradley, Databricks (databricks.com/blog/2015/03/25/topic-modeling-with-lda-mlib-meets-graphx.html)

GraphX: Graph analytics in Spartk, by Ankur Dave, Databricks (spark-summit.org/east-2015/graphx-graph-analytics-in-spark).

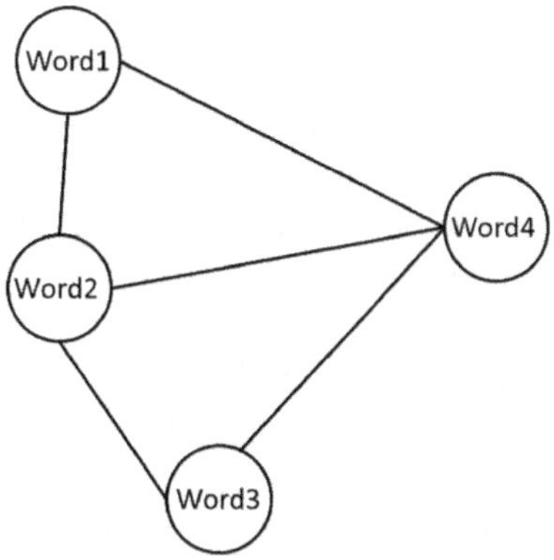

	Word1	Word2	Word3	Word4
Word1	0	1	0	1
Word2	1	0	1	1
Word3	0	1	0	1
Word4	1	1	1	0

Figure 4.7: Matrix representation of a graph model.

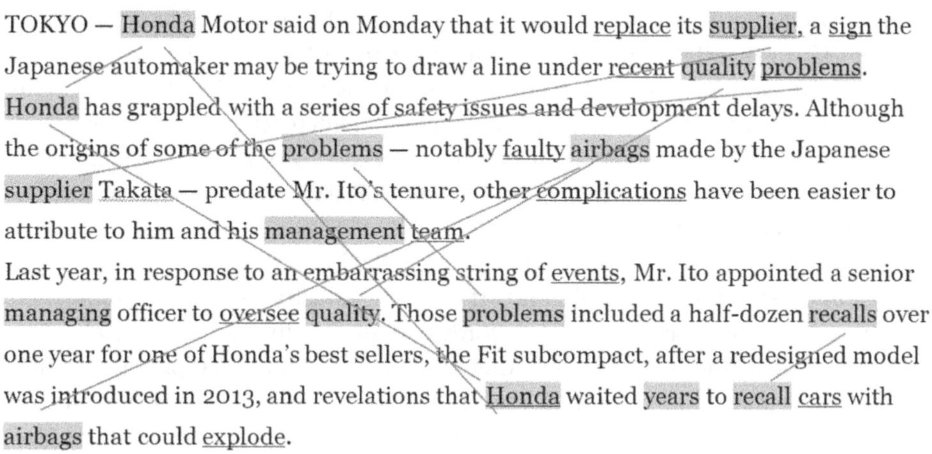

Figure 4.8: Example text analysis using graph technique.

a simple API for NLP tasks such as POS tagging, noun phrase extraction, sentiment analysis, classification, translation and more.

Pattern is another tool, a web-mining module for Python. It has tools for mining web sites (such as Google, Twitter, wikis, web crawlers, and HTML DOM parsers), NLP (POS, n-gram search, sentiment analysis, WordNet), network analysis, machine learning (clustering, support vector machines) and canvas visualization.

Additional information on NLP tools can be found online. Some of the online sources include:

- Stanford NLP Group: nlp.stanford.edu
- Carnegie Mellon University Language Technologies Institute: https://www.lti.cs.cmu.edu/work/category/2717
- University of Illinois Cognitive Computation Group: https://cogcomp.cs.illinois.edu/

4.11 Putting It All Together

To enhance understanding of textual information, we need to go beyond tokenization and part of speech tagging. Part of speech tagging helps us identify each word's place and role in the document. Is the word a verb, noun or adjective? Knowing the answer can help with understanding the subject, object and type of communication that is taking place.

Parsing text, tokenizing, lemmatizing, and even accurately assigning part of speech (POS) tagging can get us only so far in natural language processing. Understanding text requires more knowledge of context and situated cognition.

Going beyond, POS, we can apply sentiment analysis and find the polarity of sentences with some effort. Polarity of each sentence can be positive, negative, hypothetical, conditional or speculative. Combining the techniques discussed in this chapter gets us closer to computing the sentiment level from the text.

But, what constitutes understanding? Can your NLP be good enough to detect sarcasm? Can your NLP learn and adapt to new words? Computer Science Linguists and Cognitive Computer Scientists have worked on this hard problem for years. Some approaches call for using ontologies, taxonomies and inferencing. Inferencing can be difficult. Building domain expert components is a less known but effective approach to enhanced understanding. Figure 4.9 offers an NLP stack that can support increasing levels of NLP capability.

You can purchase ontologies related to the field and domain for the text you wish to understand. For example FIBO (Financial Industry Business Ontology) offers a financial and business set of keywords that are specific to the financial industry. UMLS, LOINC and ICD-10 on the other hand are ontologies that provide medical and pharmaceutical interpretations and keywords. Most organizations start by purchasing or adopting an open source taxonomy initially, but add their specific lexicon and rules to customize the taxonomy to their business needs.

To build a domain expert, you need to capture the knowledge of a subject matter expert into a set of words and concepts that describe the entities, words and their interpretations in different possible situations and scenario. Domain experts offer the context and situated cognition to enhance understanding.

Note that in the domain expert layer, I've included domain-specific libraries and rule sets. These libraries are specific to your domain of knowledge. For example, you may decide to include a set of domain knowledge about legal, marketing, financial, technology, human resources and so on. If you're in a specific line of business, say in insurance, then a domain expert component for insurance business should be developed.

Domain expert libraries can be developed manually be selecting keywords and rules that define their relationships. For example, in insurance industry the keywords "term life" and "whole life" are related, hinting that the text is likely about life insurance. For example, a rule might state that: "IF term life and age appear

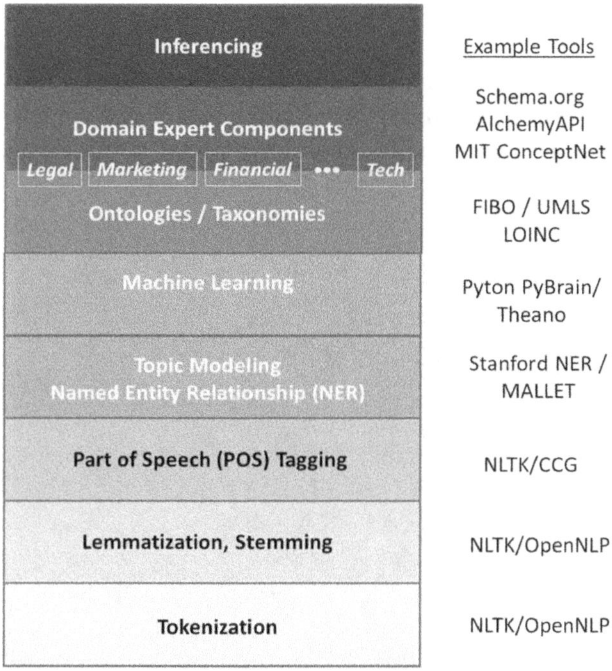

Figure 4.9: NLP stack for enhanced semantic understanding.

with in the first 50 words of a document together, the document is about life insurance," and not about other types of insurance like auto, home, etc.

You can certainly create domain knowledge using machine learning and reverse engineer the process. For example, you can take some documents, articles, news and even Wikipedia pages that are written about insurance topics and extract the keywords and their relationships automatically.

Solutions such as IBM's AlchemyAPI and MIT's ConceptNet are excellent sources to experiment and build domain expert components. Another open source tool is Schema.org. Schema.org offers semantic ontologies that capture domain specific concepts. For example, the keyword taxi, involves several associated words and conditions that are defined in Schema.org. For example, words such as destination, fare, payment type, and pick-up location take specific meaning in the context of taxi. Schema.org offers a library of semantic definitions for such keywords.

Finally, machine learning can be your ally in building, adapting and expanding ontologies. Consider ingesting a corpus of information from the financial industry sources, say for example, from CNN/Money. You can begin building your own ontologies and domain experts by identifying the frequently used keywords in such texts and then mapping them to a definition from Wikipedia pages. There are many similar "secrets" techniques in the trade that data scientists use to enhance understanding.

As we conclude this chapter, it's important to note that NLP continues to be a challenging and labor intensive task. Much of NLP techniques are still evolving but the future and vision of a day when we can understand text with high accuracy is not too far away.

Chapter 5
Quantitative Analysis—Prediction and Prognostics

Machine learning and deep learning are the latest techniques used for prediction. Prediction is often described in terms of probability or odds ratios. We predict the occurrence of an event and since future is uncertain, we offer a probability or confidence about the prediction.

The legend of Delphi gives an interesting perspective into prediction. In Greek mythology and history dating back to some 300–500 BC, Delphi was a special place. Its temple was the place where priestesses called oracles would gather and foretell the future. The oracles often entered into a state of trance (possibly after inhaling the volcanic gases in that area) and began speaking deliriously predicting one's future.

There is a legend where the Roman emperor asked the oracle if he should wage war on Persians. The oracle responded: "A great ruler will fall." The emperor took that to mean the Persian king will be defeated, so he attacked. But his army was defeated and he was killed in the battle. Similar legends tell similar stories about the assassination of Julius Cesar.[44] The lesson to be learned is twofold: First, any prediction is highly subject to interpretation. It's important that predictions are understood and considered in the proper light and context. The second lesson is that often predictions are ignore or dismissed when they don't fit our mental model. One way to apply these lessons in practice is to apply multiple predictive models in an ensemble. When we review the ensemble of models in the future chapters, I'll introduce the concept of an oracle based on this legend. The oracle described in this book, is an overseer who selects the best combination of ensembles to produce the most accurate prediction.

There's also the legend of Cassandra. She is believed to have been a beautiful goddess with the gift of prediction. But she was perceived by people to be mad. In fact, when she predicted that Trojans would enter the city in a wooden horse, the people of Troy laughed at her. This is not too far from reality of prediction today. Even the best predictive models can produce such profound and seemingly unbelievable predictions that executives are not ready to accept.

I'll leave you with one more story about prediction. The legend has it that around a thousand years ago in Persia, an astronomer lived who was excellent at predicting the future. The king was suspicious and put the astronomer to test. So he ordered his men to erect a square structure, each side with a door facing East, West,

44 Cesar was assassinated by a group of senators on March 15, On his way to visit the senators, a seer warned Cesar that he will be killed by that day, but Cesar ignored the warning. That day is remembered as the Ides of March.

https://doi.org/10.1515/9781547401567-006

South and North. He placed the astronomer in the structure and told him to predict which door will be opened for him to come out. The reward for getting the right answer was substantial amount of gold. But, the penalty of not getting the prediction right was death, a rather severe punishment for our data scientist. The astronomer made some calculations, wrote the prediction on a piece of paper, sealed it and gave it to the king.

The king ordered his men to excavation a new hole in the corner of the structure. The astronomer came out to face his fate. The king opened the sealed paper. It read: "His Majesty shall choose a 5th door for me to exit the structure."

Legends aside, prediction is both science and black art. Knowing the data modeling techniques is one thing, but knowing how to train the models and work to enhance their accuracy is labor intensive and requires a good measure of understanding signals in the data.

One approach to making predictions is done by applying machine learning to data. Predictions are valuable. They can help us prepare for an event and even prevent it. In the case of medical treatment, prevention is important in improving outcomes, more satisfied patients, avoiding preventable procedures and costs. In retail, knowing when a customer is ready to make a purchase offers a critical advantage. Knowing this information, retailers deliver offers, coupons and deals that are perfectly matched to the customer's needs at the right time and the right place, making the deal irresistible.

In the next sections, I'll review probabilities and odds ratio approach first and then review the feed forward and feedback concepts. Finally, I'll peel layers of deep learning and neural networks as we move forward to cover more technical definitions of these methods.

5.1 Probabilities and Odds Ratio

In probability, when we measure the chance that an event occurs, we divide the number of occurrences by the total number of opportunities for the event to occur. For example, the probability of getting a 6 when we roll a die is 1/6. The probability that two events occur simultaneously is multiplicative. So the probability that a player gets a double 6 when rolling two dice is (1/6)*(1/6) = 1/36.

Probability is expressed as a fraction; Odds are typically expressed as a ratio, rather than a fraction. Probability describes the fraction of time that you can expect an event to occur; Odds describe the ratio of times that the event occurs to the times that it does not. Odds of rolling 6 is once for every 5 times that you roll something else.

The odds can be from 0:1 when something never happens, to 1:0 when something always happens. Odds of 1:1 are fifty-fifty, equally likely to occur or not, which corresponds to 50% probability. A player who rolls ae die can expect to see 6 with the odds of 1 to 5. In other words, if I bet $1 to your $5 that I will roll a 6, the bets will be even in the long run. Results of logistic regression are typically expressed in odds ratios.

5.2 Additive Interaction of Predictive Variables

Often we're interested in the interaction of predictive markers and risk factors. For example, the joint effect of two predictors that estimate propensity of a customer buying a product online. Alternatively, we may want to know the extent to which the joint effect of two risk factors on a particular disease differs from the independent effects of each of the factors (Kalilani, Atashili 2006).

The joint effect is the effect of the presence of both factors on an outcome and the independent effect is the effect of each factor in the absence of the other factor. In terms of their causal effects on the incidence of an outcome, two risk factors may act independently or interact thereby augmenting (in case of synergism) or reducing (in case of antagonism) the effect of one another.

5.3 Prognostics and Prediction

This chapter offers three key ideas related to classification and prediction. First, it develops a control system treatment of prognostics and predictive models. The control system development of prognostics combines *feed-forward* and *feedback* control mechanisms to create a framework for prognostics. This framework introduces a rules-based prognostics engine that uses artificial neural network (ANN) algorithms to predict certain outcomes. In our case study, we evaluate a clinical example to predict who in a population of patients will develop a particular disease.

Second, it demonstrates the viability and feasibility of using ANN methods as predictive models in this framework. The case study explores building, training and validating four ANN models to predict medical complications from data acquired from patients' hospital stays to predict deep vein thrombosis/pulmonary embolism (DVT/PE).[45]

Third, it introduces the concept of ensemble of models. An ensemble of models (also referred to as a committee of models) can improve accuracy of prediction about future conditions about people, businesses or systems from large data sets about them. The model also provides the strength (or the impact level) of all contributing data to that prediction.

45 DVT/PE is a condition caused by blockage of patient lung vessels by blood clots that initially form in patient's legs. DVT/PE leads to severe pain, loss of lung function and even death.

The methodology enhances deep learning as it proposes using a multi-algorithm prognostics framework to enhance the accuracy of prediction. In the case study, we use four ANN models, but you can apply other models that fit the goal of your study. In this framework, I introduce a supervisory program; an overseer called an oracle[46] to select the most appropriate ensemble of models that best meet the data scientist's desired prediction accuracy.

The aim of all three ideas is to improve our ability to make predictive decisions from a vast array of data in order to be proactive and apply interventions to improve performance or to prevent negative events.

5.4 Framework for Prognostics, Prediction and Accuracy

The goal of prognostics is to offer predictions for a system's health status over a short term to long term time horizon. It's a discipline to predict future conditions in a time period ranging from a few seconds to several hours or months from the current time.

There are five critical factors that define applicability, viability and feasibility of using analytical models as prognostics tools. These factors fashioned after Smye and Clayton's work on mathematical models in medicine can be defined by the following criteria (Smye and Clayton 2002):

- Accuracy: Accuracy of prediction
- Well-posedness: Stability and immunity to small perturbations of input data
- Utility: Applicability, practicality and usability in the medical workflow
- Adaptability: Ability to handle new evidence, i.e. new data values and data types
- Economy: Cost of computation and timeliness of prediction

To explore the above critical factors in a real case study, a prognostics model using four different types of analytics algorithm are compared in the following chapters. The prediction accuracy of each model is compared to the actual clinical outcomes from prior retrospective patient cases.

Analysis about accuracy includes measurements such as calibration (agreement between predicted probability and observed outcome frequencies), discrimination (ability to distinguish between patients with and without the disease), sensitivity (true positive rate, or proportion of patients who are correctly diagnosed as having the disease), specificity (true negative rate, the proportion of healthy patients who are correctly diagnosed with a negative result), likelihood ratio (LR, the

46 Not to be confused with the company Oracle. In the historical and mythical context of Greek culture, an oracle was a mythical person who foretold the future.

likelihood that a given test result would be expected in a patient with the disease compared to the likelihood that the same result would be expected without the disease) and the receiver operating characteristic (ROC, a plot of true positive rate vs. false positive rate, namely a plot of sensitivity vs. one minus the specificity) curves (CEBM 2012).

Model validity (do the model's results match the reality), accuracy (the closeness of results to the quantity's actual value) and precision (the degree in which repeated measurements under unchanged conditions produce the same result) are compared among the four analytical models using outcomes of prior data obtained from retrospective studies.

Empirical field validation and impact analysis on prospective cases are important topics but outside the scope of this book. If you're interested in field validation you should refer to texts in clinical trial and meta-analysis that address these topics.

5.5 Significance of Predictive Analytics

As the volume of data increases, managers find it more difficult to make sense of their data and make timely decisions in their daily work. Data changes quickly and the vastness and diversity of data make it difficult for decision makers to apply all applicable decision criteria correctly at any given moment. Developing a prognostics model that can learn and adapt to specific business environments and situations is highly desirable.

5.6 Prognostics in Literature

The American Heritage Dictionary defines *prognostics* as an adjective that relates to prediction or foretelling and as a noun for a sign or symptom indicating the future course of a disease or sign or forecast of some future occurrence. Hippocrates founded the 21 axioms of prognostics some 2400 years ago (MIT 2010). The goal of prognostics is to foretell (predict) the future health (or state) of a system. A system can define a business, a market, a population, program or anything.

I present four philosophies pertaining to prediction that have evolved through the history of analytics. One is grounded in control theory. Decay of systems over time can be viewed as a result of loss of that system's control over its critical mechanisms. The goal of prediction here is to predict when a system may fail, so we can prevent a catastrophic failure or apply prevention to prevent failures. This approach requires a deep understanding of the internal causal models between control mechanisms and the system. An example of this approach includes the mathematical models using control theory that have employed differential equations to measure decay.

The second approach uses a Markov chain[47] model as it considers the cycle as a sequence of phases traversed by each subsystem from inception to retirement. For example, a customer goes through 5 stages to become a customer: suspect, prospect, lead, opportunity and customer. Consider a healthcare example: a patient with pneumonia starts from healthy, normal state and then follows four stages of congestion, red hepatization, gray hepatization, resolution (recovery). These models consider both deterministic and probabilistic approaches. The same approach can be applied to business, software, electronic systems, markets and ecology.

The third type of mathematic construct considers the asynchronous nature of the physical world and uses simulation models. For example, a study applies a simulation and statistical process control to estimate progression of changes in an economy. Another study may simulate the individual business units in a company. Simulations are models that intend to mimic the real physical world.

The fourth approach considers the system as neither a black box or a white box but as a "gray-box." Since perfect knowledge about each system's attributes and internal subsystem interactions are not available or are uncertain, one can only rely on predictive models that analyze data to make predictions.

Some of the models in this category include the survival analysis provided by the Cox hazard model and Kaplan-Meier estimate. The term survival analysis comes from biomedical research in study of mortality, or patients' survival times from the time of diagnosis of a disease to death. The first survival analysis was developed by John Graunt (Graunt 1662) who for the first time developed life tables based on his birth-death rate observations.

Survival analysis is a collection of statistical techniques used to estimate whether an event of interest will occur and at what time. Survival analysis is known by different names in different disciplines; engineering researchers refer to it as failure-time analysis; sociologists call it event history analysis while economists call it transition analysis.

Among parametric models of survival analysis the Cox hazard function is a popular method. Another method called Kaplan-Meier Estimator is a statistical tool used for non-parametric models where the mathematical equation of the system under study is unknown. Prior to these models, researchers often had to resort to life-table methods.

The Cox hazard model is a partial likelihood method that allows the researcher to estimate the regression coefficients of the proportional hazards model without the need to specify the baseline hazard function. The hazard rate is the probability that if the event in question has not already occurred, it will occur in the next time interval, divided by the length of that interval (Spruance, Reid, Grace, Samore 2004).

47 Markov chain is a technique often used for the analysis of time series data. It evaluates a series of states and the probability of transition from one state to the next.

This method is used in numerous applications including probability that a borrower may cancel an insurance policy or probability of survival after a medical procedure.

Among biomedical researchers the Kaplan Meier Estimator is the tool of choice for survival analysis. Also known as the product-limit estimator, this technique observes all data from all observations by considering survival to any point in time as a series of steps defined by the observed survival and censored times. Some models use Taylor series expansion. It's often used to measure the fraction of patients living for a certain period of time after treatment. One advantage of Kaplan Meier is in its ability to handle censored data. Censored cases are situations where a patient withdraws from a study or the certain start time of the data is not available. Kaplan Meier makes certain assumptions about data independence and uniformity that if violated can result in biased and unreliable data (Tsai, Pollock, Brownie 1999).

Other researchers have adopted a case base reasoning (CBR) method for diagnosis of medical conditions. CBR is an approach for solving a new problem by remembering a previous similar situation and by reusing information and knowledge of that information (Aamodt, Plaza 1994). Since this approach assumes that similar problems have similar solutions, it is considered an appropriate method for a practical medical domain that's focused on real cases rather than rules of knowledge to solve problems (Park, Kim, Chun 2006).

Admittedly developing mathematical models that make accurate predictions in particular in biology and medicine is challenging, but researchers suggest that soon such mathematical models will become a useful adjunct to laboratory experiment (and even clinical trials), and the provision of "in silico" models will become routine.

There are situations where collected data is inadequate to make generalizations, the evidence is not foolproof or inconclusive, and the collected data requires rapid processing and judgment.

The properties of an appropriate mathematical model for analytics include: accuracy, prediction, economy, well-posedness and utility (Smye and Clayton 2002). Among constructs used in prior research, several distinct mathematical models can be found, such as: multivariate regression analysis, Markov chains, stochastic processes, Bayesian networks, fuzzy logic, control theory, discrete event simulation, dynamic programming and neural networks.

While control theory has been widely used in systems and other engineering disciplines, it has not widely been applied to prognostics as explained in the next section.

5.7 Control Theoretic Approach to Prognostics

A system can be broadly defined as an integrated set of elements that accomplish a defined objective (INCOSE 2000). Any physical entity in the world can be considered as a system that functions as a collection of interrelated systems. The system theory approach in this book covers both feed-forward and feedback mechanisms to deliver a richer framework for prognostics.

Prognostics deal with prediction of some desired quality or characteristic of a system (Kapur 2010). It's based on understanding the science of degradation of the underlying system. The traditional systems control theory has been predominantly based on feedback: Using the feedback information from the system (also referred to as the feedback signal), a system's performance could be diagnosed, then adjusted to fix the problem. Obviously, this poses an issue. Correcting the problems after receiving feedback might be too late in certain mission critical systems and in particular when human physiology is considered.

Instead, a feed-forward model as exemplified by prognostics methods would be more desirable. Prognostics methodology is based on two principles: 1) it uses feed-forward models for prediction or to forecast of the underlying causes of problems by analyzing feed-forward signals; and 2) it suggests changes in input signals to the system in order to prevent the problem from occurring.

Prognostics and health management (PHM) is an engineering discipline that links studies of failure mechanisms to system lifecycle management (Serdar, Goebel, and Lucas 2008). Other definitions of PHM describe it as a method that permits the assessment of the reliability of a system under its actual application conditions, to determine the advent of failure, and mitigate system risks (Pecht 2008).

Within the context of systems control theory, the term "diagnostics" pertains to the detection and isolation of faults or failures. "Prognostics" is the process of predicting a future state (of reliability) based on current and historic conditions (Vichare and Pecht 2006).

Prognostics is the science of predicting the future functionality of a system by estimating the remaining useful life, probability of failure or time to failure for a given system. There are many approaches and modeling frameworks for representing a prognostics system. This is significant whether we are analyzing the failure point of business systems, an aircraft engine or human physiology.

Among many prognostics methods, computing the remaining useful life (RUL) of a system is common. RUL can be estimated from historical and operational data collected from a system. Various methods are used to determine system degradation and predict RUL.

Another method is estimating the probability of failure (POF). POF is the failure probability distribution of the system or a component. Additionally, it's common to study time to failure (TTF), the time a component is expected to fail. TTF defines the time when a system no longer meets its design specifications.

Prognostics and reliability are interrelated. Reliability is defined as the probability that a product will perform its intended function, satisfactorily for its intended life when operating under specified condition (Kapur 2010). A clinical definition can be derived from this technical description. Reliability, in the medical sense, is the probability that a patient will not develop certain medical complications during the length stay under medical care of the care provider(s). Reliability is measured by several indicators such as mean time between failure (MTBF), failure rate and percentiles of Life. Each measurement can be computed from corresponding equations that are derived from empirical and statistical distribution functions. Additional treatment of reliability can be found from text by Kapur and Lamberson (Kapur, Lamberson 1977).

Prognostics models can be classified into three general types (Eklund 2009; Hines 2009; Peysson et al. 2009). Type I is reliability based. It applies the traditional time to failure analysis by tracking a population of failures and using statistical methods for the estimation of reliability. Some typical life distributions that are used in this type of prognostics include Weibull, exponential and normal distributions. Type I prognostic methods do not incorporate the real time monitoring of operating conditions or environmental conditions.

Type II methods, also known as the stressor-based approaches consider the operational and environmental condition data. Type II considers the failures of a system in its operating environment to provide an average remaining life of a component. Some of the environmental data might include temperature, vibration, humidity and load. The proportional hazard model is an example of a type II prognostic model. Knowing the causes, one can predict reliability of a system. The simplest model in this approach is the regression model: given the operating and environmental conditions, one can predict the system failure and remaining useful life by a regression equation.

Type III prognostic methods are condition-based, namely they characterize the lifetime of a system in operation within its specific environment. They estimate the remaining life of a specific component or the entire system. Among methods used in Type III prognostics are the general path model (GPM),[48] neural network models, expert systems, Fuzzy rule-based systems, and multi-state analysis. Another example of type III approach is the cumulative damage model.

The cumulative damage model tracks the irreversible accumulation of damage in systems or components. The statistical cumulative damage model considers the number of possible damage states and a transition matrix (for representing a multi-state Markov Chain) to provide a damage prediction for multiple cyclical loads.

The goal of prognostics is to provide continuous prediction. I'll describe a case study that using an artificial neural network, and other algorithms, continuously predicts a patient's health status at regular time intervals.

48 GPM (general path model) is a technique of taking several observations of a system parameter over time and observing the degradation (or deterioration) of the parameter to predict when a system is likely to fail. Consider for example, an airplane that experiences multiple take-off and landing episodes. Each episode is an observation of the plane's metal fatigue that can lead to the plane's physical disintegration.

5.8 Artificial Neural Networks

Neural networks have been used successfully to predict future onset of events. For example predicting diseases such as recurrence of various types of cancer, cardiac illnesses and to assist physicians with prognostic and decision support. These studies have offered long term predictions for patient health conditions, typically forecasting the disease-free or disease recurrence in the future ranging from a few months to several years.

Artificial neural networks (ANNs) are parallel computational methods by interconnecting artificial neurons. They're ideal for solving non-linear problems that come with a long list and diverse types of input variables. ANNs are adaptive to specific problems and can be trained for pattern matching or classification. An ANN model can be trained by mapping a disease to a known set of input clinical measurements and then later be applied to a new patient. The trained model can match the input measurements of the patient to presence or absence of a disease. The model can even classify the patient's clinical measurements into various stages of a disease.

Since early 1980s, artificial neural networks have been applied successfully to several prediction problems in business, engineering and medicine (Delen 2009). One of the most popular models is the multi-layer perceptron (MLP) with back propagation, essentially a supervised learning algorithm. It's been shown that ANN models using the MLP algorithm are capable of learning arbitrarily complex non-linear functions to arbitrary accurate levels (Hornik, Stinchcombe, White 1989). The MLP is essentially a collection of non-linear neurons connected together by weighted links in a feed-forward multi-layer structure.

Among highly accurate models, *Support Vector Machines* (SVMs) have been proposed (Delen 2009). The SVM algorithm is not regarded as an artificial neural network model, but it's been included in the ensemble for its solver capability. Most recent algorithms use Levenberg-Marquardt methods proposed by Levenberg and Marquardt that are highly accurate as well as computationally fast ANN models (Wilamowski and Chen 1999). Support Vector Machines are not regarded as neural networks, but they can be used as a solver method in a neural network model.

Medical research has shown that certain life-threatening conditions exhibit early indicators in physiological data. A study conducted on improving neonatal intensive care units (NICU) (Blount 2010) provided interpretations of multiple streams of clinical and physiological data to detect medically significant conditions that precede the onset of medical complications for neonatal patients.

In another study (Webber 1994), A neural network model was trained on EEG data. The input consisted of 49 channels of real time EEG data to detect epilepsy spikes. The study showed that ANNs offer a practical solution for automated detection of real time epileptiform discharges using inexpensive computers.

ANNs have been shown to be a valuable tool to the clinical diagnosis of myocardial infarction (Baxt 1991). The model used in one study was trained on 351 patients admitted for high likelihood of having myocardial infarction. It was prospectively tested on 331 consecutive patients presenting to the ED department with anterior chest pain. The network was able to distinguish patients with from those without acute myocardial infarction at a slightly higher sensitivity than physicians' diagnosis for those patients.

In another study of patients in intensive care units (ICUs), an ANN model was shown to be more effective than a logistic regression model for predicting outcomes of care (Dybowski et al. 1996). The ANN model was applied in the clinical setting of systemic inflammatory response syndrome and hemodynamic shock on 258 patients. The outcome evaluated was death during that hospital admission. The best performing ANN model was trained after 7 training iterations.

In cancer treatment cases, ANNs have become a popular tool for predicting outcomes (Dayhoff and DeLeo 2001). At one institution (Bottaci et al. 1997) six different ANN models were developed to predict outcomes of individual patients who were diagnosed with colorectal cancer to predict death within 9, 12, 15, 18, 21, and 24 months. Results showed that ANNs were able to detect outcomes more accurately than the then available clinicopathological methods.

Another research project conducted with breast cancer patients (Ravdin and Clark 1992) suggested that ANNs can be trained to recognize patients with high and low risk of recurrent disease and death. Moreover, their study showed that by coding time as one of the prognostic variables, an ANN can be used to predict patient outcome over time. In particular ANN models can make a series of predictions about probability of relapse at different times of follow up, allowing clinicians to draw survival probability curves for individual patients.

In another study, a set of patients' mammography tests were interpreted by radiologists and by an ANN model (Floyd et al. 1994). The model was more accurate in detecting breast cancer patients than radiologists. A more comprehensive overview of application of neural networks in decision support of cancer found 396 studies and found that overall ANNs add better diagnostic accuracy than traditional in the field of cancer (Lisboa and Taktak 2006).

Several studies have compared the accuracy of ANNs with logistic regression models, but after a meta-analysis the conclusions are mixed. Some papers (Delen 2009) show that ANNs are far more accurate, but a few papers find both methods comparable for medical prediction (Adams, Wert 2005). The next chapter reviews the validation and viability measurements that are used as criteria to select the most appropriate model.

Chapter 6
Advanced Analytics and Predictive Modeling

6.1 History of Predictive Methods and Prognostics

A vast majority of mathematical models have been used for prediction. But, models to make predictions using prognostics have not been fully explored. Among those that did address predictive analytics, many required and made assumptions about the type of data and distribution. For example, many studies assumed normality in their data set. Most have relied on a single model to make predictions. Little attention has been paid to developing a framework for prediction of using prognostic methods on these large data sets. In this chapter, I cover several diverse techniques to predictive analytics, and in Chapter 7, I will cover a predictive modeling framework that explores multi-model predictive approach.

Traditionally, the two most commonly used data mining techniques are linear discriminant analysis (LDA) and logistic regression to construct classification models. Research in data analysis has demonstrated that generally neural networks and deep learning models[49] are more accurate than linear logistic regression models in prediction. For example in cases of new or recurrent cancer prediction (Delen 2009).

A new topic in machine learning called deep learning relies on neural network models. From a survey of literature from the 1970s to present, some of the most successful predictive methods in literature are model-free approaches using neural networks and fuzzy sets (Kodell et al. 2009, and Arthi and Tamilarasi 2008).

Logistic regression, on the other hand is a special form of linear regression models that allows non-numeric input variables (namely allows categorical data). It's used for classification or prediction with the probability that an event will occur by fitting data to a logit function (or a natural log function) logistic curve.

Logistic regression (LR) is a generalization of linear regression. It's used as the means to predict binary or multi-class dependent variables. Since the response variable is discrete, it cannot be modeled directly by linear regression. Instead, logistic regression rather than predicting a point estimate of the event itself, builds the model to predict the odds of its occurrence. When predicting the occurrence or non-occurrence of an event, it is basically a two-class problem, if the odds are greater than 50%, it implies that the case is assigned to the class designated as 1, otherwise as 0.

49 Deep learning models are a form of neural network models but with more hidden layers. The higher number of hidden layers improves machine learning from data and hence produces better results.

https://doi.org/10.1515/9781547401567-007

Logistic regression assumes that the response variable (the log of odds) is linear in the coefficients of the predictor variables. In addition, LR models do not select the best inputs and the modeler must select the right inputs and specify their relationship to the response variable (Delen 2009). These are among limitations to logistic regression. To overcome these limitations data scientists apply non-linear logistic regression models. Selecting the right input variables is also key to achieving accuracy in logistic regression models. One technique called feature selection is used to select the right variables. Feature selection can be done by running the regression model several times, each time with a single variable and observe the error term. The variables that present the lowest error term are the selected features or the significant variables that can be used in forming the final regression model.

Classification trees (also referred to as decision trees) are also used for prediction. Data scientists typically prefer classification trees over the traditional statistical tools when assumptions about data distribution cannot be met. For example, many traditional statistical methods assume normal distribution of data which in real world is not always possible.

They form hierarchical models of data with branches in a tree-like structure that lead to specific diagnoses. The pitfalls with these methods are shown to be related to their linear approach and errors associated with initial choice of data that form the tree classification.

Historically, most predictive methods have relied on either a single model for prediction and/or on linear methods for prediction. While predictive models using ANN have been reported in the literature, such models provide long term predictions that span over several years into the future, and do not focus on short term predictions.

Prior research has predominantly limited their model validation to a few criteria, such as *sensitivity*, *specificity* and *ROC* calculations.[50] Traditionally, data scientists have limited their models to only one ANN model or to the algorithm that produced the best results. They may neglect to compare results of multiple models along with these and other validation measurements. The framework prescribed in Chapter 7 shows how multi-model approaches can achieve higher accuracy, precision and performance.

Before building models, it's important to think about validation and viability of the model. In other words, how would we test our model for accuracy and how will be able to validate its results? The next section presents some challenges, pitfalls and techniques to overcome them.

50 Receiver Operating Characteristic (ROC) curve.

6.2 Model Viability and Validation Methods

Model validation and verification are important steps for providing confidence and credibility in the model's results. Verification (ensures that the model performs as intended) and validation (ensures that the model represents and correctly reproduces the behaviors of the real world system) are essential elements of model development for practical applications (Macal 2005). Model verification deals with building the model right. Validations deals with building the right model.

The goal of validation is to ensure that the model addresses the right problem, provided accurate information about the system being modeled. One of the dangers to modeling validity is "overfitting" the model to a given data set, where the model is fitted to a specific dataset. Overfitting can occur when important elements of the model reflect randomness in the data rather than underlying model drivers. To overcome this limitation, researchers have employed techniques such as cross-validation, or keeping a "hold-out" random data sample to perform testing on a separate data set. In addition, researchers have considered accuracy measures of the model using prior data as the expected results.

Three aspects of validity that must be followed are:

1. Calibration – agreement between observed probabilities and predicted probabilities
2. Discrimination – the ability of the model to distinguish between different outcomes
3. Usefulness – the ability of the model to improve the decision making process

There are two types of validation, internal validation and external validation. Internal validation uses techniques such as cross validation and boot-strapping to assess the performance in samples of the same population. External validation is the process of measuring the performance of the prediction model in samples from different populations such as patients from other locations.

With the cross-validation technique, the model is developed using a randomly selected part of the data sample and tested on the rest of the sample, then the process is repeated several times and the average is computed as an estimate of performance. With the boot-strapping technique, a sample of the same size as the development sample is randomly selected with replacement, the model then is developed using the boot-strap samples and tested on those not included in the boot-strap samples.

Tests that measure clinical usefulness include accuracy, sensitivity, specificity and decrease in weighted false classifications, namely the number of false negatives and false positives. Tests that measure discrimination include the receiver operating characteristic (ROC) curve, a plot of model *sensitivity* vs. *(1–specificity)*. Finally tests that measure calibration include a calibration plot and the average absolute different between observed frequencies and predicted probabilities. These measurements are defined and employed in this framework as described in the upcoming chapters.

External validation requires real world tests such as A-B testing, clinical trials and specific design of experiments that are outside the scope of this book. A-B testing measure the actual results of a prediction or decision between two groups: Group A that receives the decision and group B that serves as the control group.

6.3 Classification Methods

Statistical methods include algorithms that identify trends, measure expected value, and identify outliers and variability. Often a regression line provides the basis for determining the best line that fits a number of data points in a scatter diagram. Given the regression line, one of the techniques to identify outliers is to define an acceptable range above and below the regression line, forming an acceptable band. These upper and lower lines define trim-points. All data points that fall outside of this band (outside of trim-points) are regarded as outliers.

6.4 Traditional Analysis Methods vs. Advanced Analytics Methods

The traditional analysis methods were fundamentally hypothesis driven. Researchers devised a null hypothesis and using statistical methods worked to show the null hypothesis can be rejected. The hypothesis was devised based on a clue about interaction and causality between data elements. The null hypothesis is a general statement that there is no relation between two data variables (the two phenomena are independent).

The goal of hypothesis testing is to statistically reject the null hypothesis with a low error rate. The error rate is shown by alpha and a strong test shows that the hypothesis can be rejected with 95% confidence and only 5% chance that the data can't reject the hypothesis due to error. The error rate should be as small as possible and typically it's preferred to be under 5% ($\alpha < 0.05$).

The new data analytics methods make no prior assumptions about data relationships. They view data as the model itself. As one data scientist put it: "The data is the model." In new data analytics methods, models evolve. They're not static. Models adapt with new data as new data becomes available. The data analytics methods look for hidden patterns or relationships in data elements. They work best when data elements are not highly correlated. They consider outliers as being important to the data analysis. They find patterns that point to new insights which can potentially form grounds for a new hypothesis. In general, data analytics methods help us ask questions that we're not aware of.

6.5 Traditional Analysis Overview: Quantitative Methods

The traditional analytical methods are commonly found in research papers and publications. These methods include:

- Time-series: Observing a single variable over time. A line-chart can be used to demonstrate the trend.
- Ranking: Ranking data in ascending or descending order. Creating sub-divisions in data by ranking it (low/med/hi).
- Ratios: Ratios can show the relationship between two variables (return on investment : ROI), or as a ratio to the whole (a quarter of population: 25%). A pie chart or bar chart can be used to show the result.
- Deviation: Data is compared to a reference point, such as comparing actual vs. budget expenses. A bar chart can show the deviation between the two values.
- Frequency distribution: The number of observations of a variable is shown as the how frequently the variable is observed at certain values. For example, the deposits under \$1000, between \$1000 to \$50,000 and over \$50,000. You can use a bar chart of histogram to show this data.
- Correlation: Shows how two variables compare as they increase or decrease in value. If you plot revenue and customers, you get a positive correlation. The more customers you have the higher the revenues. The opposite is a negative correlation. If you plot supply and price curves, there is a negative correlation in that as supply increases, prices decrease.
- Nominal comparison: Analyzes values of variables for nominal (or categorical) sub-divisions of data. Example: Sales volume by month. A bar chart is used for this type of analysis.
- Geospatial: Comparing two or more variables on a map. Example: Sales by region or by city. A cartogram is used to illustrate the result.

This chapter starts our review of predictive approaches by covering simplest and easiest methods like regression analysis and builds towards more complex techniques.

6.6 Regression Analysis Overview

Regression is generally expressed in terms of a dependent variable Y, a set of one or more independent variables X and coefficients of these variables. The equation in brief form is expressed as:

$$Y = b + wX;$$

Where Y is the dependent variable, b is known as the intercept and w represents the coefficients. We can draw this equation in two dimensions as a straight line where b is a constant term and w represents the slope. We shall see later that this regression equation is a simplification of methods used for data clustering and data classification.

Imagine a scenario where a property valuation (home appraisal) is needed. The home price is computed based on the price of homes sold in an area (dependent variable) and the characteristics of each home (independent variables) such as home size, number of bedrooms, bathrooms, garage and age of the home. The price is likely to be higher with larger property size and number of bedrooms and bathrooms. Let's assume that in this location, the older the home, the lower the price. The equation might offer the following coefficient structure in this scenario:

$$P = 0.3s + 0.45b - 0.1a$$

Where P is price, s is the home size in square feet, b is the number of bedrooms, and a is the age of the home.

Note that in this equation the regression coefficients represent the independent contributions of each independent variable to the prediction of the dependent variable. This is to say that variable X is correlated with the variable Y, after controlling for all other independent variables. This type of correlation is known as a *partial correlation*. A partial correlation between two variables is achieved when we remove the effect of all control variables that might have a confounding effect on the dependent variable.

To understand the correlation between variables, look at the signs (plus or minus) of the regression coefficients. If a coefficient is positive the relationship between this variable with the dependent variable is positive. For example, the higher the education level, the higher the salary). If a coefficient is negative, the relationship between this variable and the dependent variable is negative. For example, the smaller the class size, the higher the test scores).

In our home value appraisal scenario the home size, number of bedrooms and baths are positively correlated with price, but the age of the home is negatively correlated. In the regression equation, the sign of each independent variable indicates the direction of correlation. We should not draw any conclusions about the strength or relative importance of the coefficients. For example, it would be incorrect to assume that the property size is 3 times more important than the age of the home in computing the price of a home, as the scale of these variables could vary considerably.

Regression results are the best estimates possible, but the actual values will be different because there are some other variables that we've not accounted for in our model. The difference between results of the regression equation and actual observed values are called residual values. A residual value is the deviation of a particular point from the regression line (its predicted value).

A measure of error is called R squared, also known as the coefficient of determination. R squared is 1 minus the ratio of residual variability. The R-squared value varies between 0 to 1. When the regression model is a poor fit (where there is no relationship between X and Y) then R-squared will be zero. If Y and X, are perfectly related then there are no residual variance between the regression points and actual values, then R-squares will be 1. So, a higher R-squared value is more desirable.

For example, if R-squared equals 0.60, it implies that we've accounted for 60% of the original variability and only (1–0.60) or 40% is due to residual variance. In other words, the R-squared value is an indicator of how well the model fits the data.

In the context of the regression equation, we can think of the independent variables (the X variables) as predictors. The degree to which predictors (X variables) are related to the dependent variable (Y), is shown by the correlation coefficient R, which is the square root of R-squared.

One of the biggest assumptions about multiple regression is linearity. It's important that you check the data in advance to ensure it is linear. Typically a scatter plot of data will reveal non-linearity. There are other techniques to test the data such as non-linear models, general stepwise regression, exploratory data analysis and data mining techniques which we'll cover in the following sections.

The other assumption is that of a normal distribution. Regression models assume that the data follows a normal distribution. You should check for normality using statistical tests or a scatterplot diagram to visually inspect the data distribution.

Another noteworthy fact is that while regression explains relationships between variables, it should not be used to explain causality. Just because two variables are correlated, we cannot conclude that one causes another. The causality analysis is more complex and difficult to prove. Causal analysis covered in the future sections addresses this topic.

It's easy to get carried away and include too many variables (predictors) in the multiple regression just because it's easy to do. However, you must have a reasonable number of data points (observations) to build a sensible regression model. If too many predictor variables are selected, the model becomes unstable and unlikely to replicate if you were to conduct the study again. The rule of thumb in the industry and common practice among data scientists is that you should have 10–20 times as many observations as you have predictor variables.

6.7 Cox Hazard Model

The Cox Hazard model has many applications ranging from insurance predictions (such as insurance policy change) to predicting patient disease condition in healthcare. It uses the Cox regression equation. The model defines the relationship between survival of subjects (for example in healthcare it means patients, in insurance, it means policy holders) and several explanatory variables. Survival analysis, as we have mentioned before, is concerned with studying the time between entry to a study and a subsequent event (such as death).

A Cox model provides an estimate of the treatment effect on survival after adjustment for other explanatory variables. In addition, it allows us to estimate the hazard (or risk) of death for an individual, given their prognostic variables.

The Cox regression model is essentially a regression equation shown as:

$$y = b + w_1x_1 + w_2x_2 + \ldots + w_kx_k$$

A positive regression coefficient for an explanatory variable means that the hazard is higher, and thus the prognosis worse. Conversely, a negative regression coefficient implies a better prognosis for patients with higher values of that variable.

Another important and commonly used analysis method is correlation analysis.

6.8 Correlation Analysis

The purpose of correlation is to find relationships between variables. The most commonly used correlation coefficient is *Pearson r*, a simple linear correlation method, also known as linear or product –moment correlation. It determines the extent to which two variables are proportional to each other. Proportional means that the two variables have a linear relationship and this relationship can be represented by a line, called the regression line. We've seen what the regression line looks like and know how to compute it. The correlation coefficient (r) represents the measure of the relationship between two variables. When the correlation coefficient is squared (r^2), it will represent the proportion of common variation between two variables, namely the strength or magnitude of the relationship.

Another important measure is the *significance of correlation*. The significance of correlation determines the reliability of correlation. When the sample size is larger, we can expect the reliability of correlation to increase. The test of significance is based on the assumption that the residual values follow a normal distribution and that the variability of the residual values is the same for all values of the independent variable x.

Researchers use a rule of thumb that if your sample size is 50 or more, then bias is not a big concern. For sample sizes of over 100, the assumptions in particular the normality assumption should not be a concern.

Let's review these topics once again:
- Correlation is the measure of relation between two or more variables.
  - Correlation coefficients range from –1.00 to +1.00 (perfect negative and perfect positive correlations)
  - Most common correlation measure: Pearson's *r*

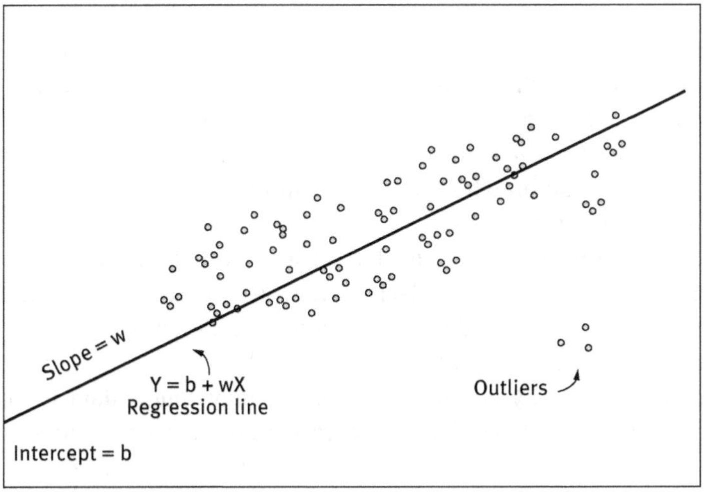

Figure 6.1: Graphical representation of regression equation in 2 dimensions.

  - The correlation is a linear relationship shown as a regression line (see Figure 6.1)
- *R squared* (r^2) is the proportion of common variation between two variables[51]
- Significance of correlation indicates reliability of the correlation
  - A significance test assumes a normal distribution of residual values
  - Larger sample sizes of 50 or more reduce bias concerns

Outliers are atypical, infrequent observations. Outliers can significantly skew the slope of a regression line and thus the value of coefficients. One should not blindly run a correlation analysis without considering outliers. Since the regression line is determined by minimizing the sum of square of distances of a data point from the line (not just the sum of simple distances), outliers have a huge impact on the slope of regression line. An outlier can change the coefficients to be higher or show a higher correlation in a study that the value of correlation (without the outlier) would be near zero.

A scatterplot is good way to spot outliers. Researchers often exclude outliers by removing data that is more than ±2 Standard deviations from the mean. In many cases, outlier detection is an important way to detect new insights about data. Outliers may signal anomalies or black swan events.

51 R-squared represents the proportion of variance of the dependent variable that's explained by independent variables in a regression model. Investopedia defines this relation in investing as the percentage of a fund that can be explained by movements in a benchmark index.

Black swan events are based on the philosophy that most people who have only seen white swans believe that swans are always white. In the southern hemisphere, the observer finds a black swan which fundamentally changes the observer's beliefs. Rare events that are extreme outliers are known as black swan events.

Outliers can significantly skew the slope of the regression line and coefficients. Outliers may result in high correlation when correlation would be near zero. But, a scatterplot is good way to spot outliers.

As researchers we want to remove or reduce bias. There are 12 types of bias to be watchful of. One common source of bias is lack of homogeneity in data that can cause bias. For example: a researcher conducts two studies at different times or geographies and combines the data from both studies.

Correlation assumes homogeneity of data. A researcher may collect data from two separate population groups, or two different geographies or even time periods. The data "groups" should not be combined as shown in the diagram. The scatterplot shows how observations cluster into separate "clouds." The combined data results in a false or biased regression line that shows a high positive correlation. But, this is only due to the way the two data sets are positioned. Figure 6.2 illustrates the effect of non-homogeneity of data on correlation.

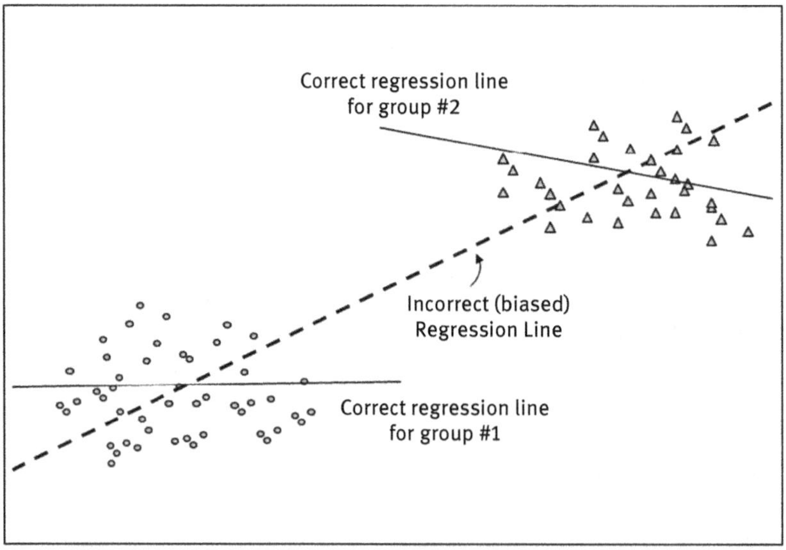

Figure 6.2: Effect of non-homogeneity on correlation.

When you compute a separate regression line for each data set, you might find surprisingly different results, as shown in this scatter plot. The correlation for group#1 is near zero and the correlation for group#2 is negatively correlated. Working with homogeneous data is important to achieve correct correlation analysis.

6.9 Non-linear Correlation

Many of the interactions we study in nature are non-linear. Almost in every industry and discipline like biology, economics, retail, transportation, finance and in particular in medicine, we often find non-linear relationships among data. As we reviewed earlier, *Pearson's r* assumes linearity in data and measures the linear correlation between data values. But, what if our data is non-linear? A scatterplot is a good method to visually detect non-linearity. An example is shown in Figure 6.3.

If the data is monotonic, we can apply a non-parametric correlation method such as *Spearman R* which is a technique sensitive to the ordinal arrangement of values. The other approach is to use the "goodness of fit" functions such as spline functions. The other technique is to divide the data into 3 or more equal width segments and treat the segmentation as a new grouping variable and run an analysis of variance on the data. We'll cover analysis of variance in the next sections.

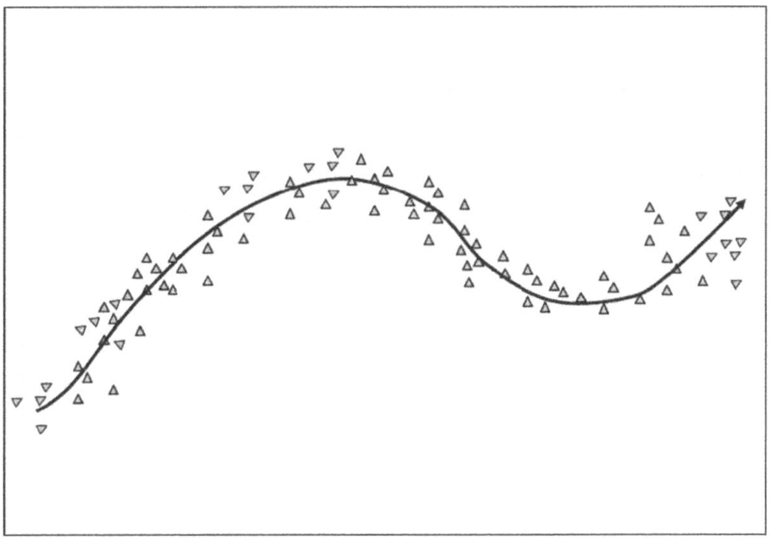

Figure 6.3: Scatter plot of a non-linear data set.

6.10 Kaplan-Meier Estimate of Survival Function

The Kaplan-Meier method determines the survival rate of a population under study. It measures the survival rate beyond the duration of a study. It computes the proportion of people who survive the length of the study. The study in insurance may predict the number of people who remain loyal customers, or the number of people who will file a claim. This is a form of time-series analysis where data is observed at certain intervals. In healthcare, Kaplan-Meier method is commonly used to predict the impact of a treatment on patient's survival rates. Consider an example where 10 patients are selected for a study. The method measures the number of deaths and censored patients.

When a subject leaves the study, we call that a censored subject and reduce the number of patients by one. Generally, we measure survival rate at any given time as the ratio of live patients entering the study minus the patients who expire, divided by the number of patients living at the start of the study.

S = (no. of live subjects entering the study – no. of subjects died) / no. of live subjects entering the study

Subjects who leave the study are censored and not counted in the denominator. Total survival rate at a given time interval is calculated by multiplying all probabilities of survival at all time-intervals preceding that time interval. Figure 6.4 illustrates an example of the Kaplan-Meier calculation.

Table 1. Calculation of Kaplan–Meier estimate of the survivor function

A Survival time (years)	B Number at risk at start of study	C Number of deaths	D Number censored	E Proportion surviving until end of interval	F Cumulative proportion surviving
0.909	10	1	0	1 – 1/10 = 0.900	0.900
1.112	9	1	0	1 – 1/9 = 0.889	0.800
1.322*	8	0	1	1 – 0/8 = 1.000	0.800
1.328	7	1	0	1 – 1/7 = 0.857	0.686
1.536	6	1	0	1 – 1/6 = 0.833	0.571
2.713	5	1	0	1 – 1/5 = 0.800	0.457
2.741*	4	0	1	1 – 0/4 = 1.000	0.457
2.743	3	1	0	1 – 1/3 = 0.667	0.305
3.524*	2	0	1	1 – 0/2 = 1.000	0.305
4.079*	1	0	1	1 – 0/1 = 1.000	0.305

* Indicates a censored survival time

Figure 6.4: Kaplan-Meier estimate for 10 subjects.

Then the calculation for each row follows the steps below:

- In the first year, one death has been observed and no patient was censored. So the proportion surviving changes from 1.00 (or 100%) to (1–1/10) = 0.900. We use the same figure as the cumulative proportion surviving the first year of the study.
- In the 2nd year, notice one more patient has died. The proportion surviving until the end of the interval is (1–1/9) = 0.889. Next multiply 0.889 by 0.900 to get 0.800. This is the new cumulative proportion surviving.

- In the 3rd year, there are no deaths but one censored patient. The proportion surviving is (1-0/8) = 1.000. Next we multiply 1.000 by 0.800 to get 0.800. This is the new cumulative proportion surviving in the third year.
- In the 4th year, there is one patient death, so the proportion surviving is (1-1/7) = 0.857. Next multiply 0.857 by 0.800 to get 0.686. This is the cumulative proportion surviving the 4th year.
- We continue to this calculation for every year until year 10, when the cumulative proportion surviving comes to 0.305, namely 30.5% of patients who enter into the study are expected to survive the 10-year study.

6.11 Handling Dirty, Noisy and Missing Data

Missing data can pose two issues for data scientists. It can introduce bias due to the absence of data for certain variable and also loss of precision due to reduced sample size. A traditional approach for handling missing data is the last observed carried forward (LOCF) method. LOCF method is often used when there is missing data in randomized control trial studies. It works by using the last observed data to fill in the missing values. This is called the LOCF imputation method. Imputation is he practice of substituting a value for the missing data. The LOCF imputation method is used when data is longitudinal. This method requires knowing the data structure and data record sequence, and the time the data was recorded. Figure 6.5 shows an example of an LOCF imputation. The last observed values (such as 2.0 for the first row, and 3.5 for the third row) are carried forward.

	Observation time						
Unit	1	2	3	4	5	6	...
1	3.8	3.1	2.0	? –> 2.0	? –> 2.0	? –> 2.0	
2	4.1	3.5	3.8	2.4	2.8	3.0	
3	2.7	2.4	2.9	3.5	? –> 3.5	? –> 3.5	

Figure 6.5: An illustration of LOCF method.

While the LOCF method is very easy to apply for full longitudinal data analyses, it has some downsides. For example, it may distort means and covariance structure. The National Academy of Science Advisory recommends those imputation methods that provide valid Type I error rates. We'll cover this recommendation in more detail in the next sections.

Some other options for imputing missing data include:

- Assume all missing data were poor outcomes (this has its disadvantages, but assumes worst case scenario)

- Impute by placing the missing value by the mean or median value (disadvantage: weakens covariance and correlation estimates)
- Impute by predicting the missing value using the regression equation (disadvantage: overestimates model fit and correlation estimates and weakens variance)
- Create a dummy variable: Create an indicator for missing value (1=value is for a missing observation; 0=value is for observed value); impute the missing value (by mean or regression); then include the missing indicator in the regression model (disadvantage: can produce biased results)
- Impute by using statistical models such as maximum likelihood estimation and multiple imputation (explained in the next section). In Chapter 3, we read about a case where missing data was imputed using K-mean clustering technique. K-mean approach considers imputing the missing value to considering the relationship of the missing value with other data elements in the data set.
- Bootstrap methods: use methods that provide valid Type-I error rates like the bootstrap & generalized estimation equation (GEE). The bootstrap method uses random resampling with replacement and allows assigning measures of accuracy of bias, variance or Type I error
- Maximum likelihood estimation (MLE): MLE identifies the set of parameter values that produces the highest log-likelihood, in other words the value that is most likely to have resulted in the observed data. There is a method called "PROCMI" in SAS that runs MLE and produces this value.
- Multiple imputation method (MIM): Missing data is filled with repeated imputed values from multiple regression models. It results in better imputation values with more accurate variability. In SAS and STATA there are models like "ICE" (Imputation using chained equations) and "MIM" (multiply imputed model[52]) data set.

The basic idea behind maximum likelihood estimation is to estimate a value that is most likely to the right replacement for a missing value. Assume you have a set of variables x1, . . ., xn, and you're missing one of the values among the set. You make an assumption about the distribution of the data, such as normal distribution, Bernoulli distribution, or binomial distribution, or other type of distribution. Given the type of distribution, you can measure the mean and standard deviation of the data set. Using the mean and standard deviation, you can define an estimation function and estimate the missing value.

The multiple imputation method uses multiple regression to estimate the missing value. Multiple regression uses existing known values in a regression equation

52 MiM and ICE are functions available in STATA to conduct multiple imputation on datasets with missing data.

to estimate the value of the missing variable. It was coined by Pearson in 1908. See the content under regression.

6.12 Data Cleansing Techniques

Despite best efforts by researchers to deliver valid research results, errors and mistakes occur. *Good clinical practice* methods have been established to minimize error in study methods and data handling. The most recent emphasis from good clinical practice guidelines focus on standards for documentation, data cleansing, data handling methods. In addition, the Society for Clinical Data Management has a guideline for minimum acceptable data quality levels for clinical trials. Data cleansing has traditionally been viewed as a suspicious activity, bordering on data manipulation.

More recently, there has been more emphasis on data cleansing methods to distinguish it from data manipulation. Statistical societies recommend that the description of data cleansing be a standard part of reporting statistical methods. Researchers Jan Van den Broeck et al. offer a 3-stage data cleansing process, "involving repeated cycles of screening, diagnosing and editing of suspected data abnormalities." ("Data Cleaning: Detecting, Diagnosing, and Editing Data Abnormalities, PLOS Medicine, Oct. 2, 2005; Jan Van den Broeck, Solveig Argeseanu Cunningham, Roger Eeckels and Kobus Herbst).

The screening process involves detecting four basic types of abnormalities:
- A lack or excess of data
- Outliers and inconsistencies
- Strange patterns including joint distributions
- Unexpected analysis results

One form of data error is erroneous inliers. Erroneous inliers are data that fall into the valid range but are generated by error. In order to detect these inliers, we must examine the history of each data point, use scatter plots, regression analysis or consistency checks or even re-measure a sample of inliers to estimate an error rate.

The *diagnosis phase* works to determine what is wrong with the data. It looks to identify issues with data, missing data, erroneous data and extreme data points. It checks data within the range of cut-offs set by the researcher. The low and high end of the range may be assigned soft cut-off and hard cut-off values. Data within the soft cut-off is regarded as suspect, to be further examined. Data values beyond hard cut-off are regarded as obvious errors and must be treated and fixed. In longitudinal studies, the flow of data and temporal measurements are expected. We can detect outliers and errors by considering the data flow or the expected temporal recordings.

The *treatment phase* works to deal with the erroneous data. The researcher must determine how to treat the data. The options are limited, either remove it, fix it or leave it alone. A researcher may conduct a re-measurement of the study if it's within a short time span from the initial measurement. The researcher may take the average of the initial and new measurements as final data treatment. The researcher may opt to remove the erroneous data but consider the impact of removing the data and report it as excluded from analysis.

Finding erroneous data can help as feedback into the data collection process and improve future designs of research programs.

Let's review the four stages of data cleansing in more detail:

- Screening phase
  - Detect four basic types of abnormalities: In the screening phase, look for lack of data or excess data that is not commensurate with the scope of the study. You want to look for outliers and inconsistencies in data. Also, you want to identify strange patterns including joint distributions and unexpected analysis results.
- Methods for screening:
  - Look out for erroneous inliers, examine the history of each data point, use scatter plots, perform regression analysis, conduct consistency checks and as a last resort re-measure a sample of inliers to estimate an error rate for your data.
- Diagnosis phase:
  - Look to determine if the data is erroneous? Examine the data to determine what is the issue and cause of the erroneous data? You should ask do we have missing data, or extreme data points?
  - The diagnosis phase checks data within the range of cut-offs set by the researcher. You should examine the low end and high end of the range with soft cut-off and hard cut-off values. Data within the soft cut-off is regarded as suspect, and should be further examined. Data values beyond hard cut-off are regarded as obvious errors and must be treated. In longitudinal studies, the flow of data and temporal measurements are expected. We can detect outliers and errors by considering the data flow or the expected temporal recordings in a study. Data points that fall outside of expected data flow or time intervals should be regarded as suspect.
- Treatment phase:
  - Data cleansing (data scrubbing) removes invalid data points from a data set. You may delete data that does not fit the data series, pattern or frequency distribution. You must apply data transformations to de-duplicate data. But, first you should test data for finding matching data records to identify duplicate records.
  - Do not summarily remove outlier data. Be careful about deleting outlier data that may have significance. Cleansing (deleting) data can be done by human judgment if source or data collection processes are not trusted. Constraint tests can detect inaccurate data (e.g. SSN No: 999-99-9999),

data out of range, format mismatch, and foreign key checks. You can detect issues such as missing data, "0" when blank or N/A was expected, "999" or "9999" indicating no data in your data set.

Let's review this topic in more detail. Data cleansing, data scrubbing or data cleaning is the process of removing invalid data points from a dataset. Its goal is to delete data that does not fit the data series, the assumptions about the data, a pattern or frequency distribution for that data.

Our goal is to identify dirty data—incorrect, incomplete, invalid and irrelevant data—and replace with good data. Data cleansing and scrubbing involves de-duplication of data and removing data. In order to re-duplicate data we need to match records first to identify duplicates. The other approach is to identify outliers and remove them.

But, care must be applied when removing outlier data that may have significance. Ensure that the outlier is truly "bad" data. Some outliers are significant as they might indicate additional clusters of data points that should not be ignored. Extreme outliers can be helpful as they inform us of Black swan events.

Cleansing (deleting) data can be done by human judgment. For example, if the data source or data collection process are not trusted, the researcher may remove that data manually without statistical tools.

Detecting invalid data can be done algorithmically through a number of techniques by applying constraints such as:

- Check the value with in a range of data. For example, a negative value for cholesterol level or a value over 130 for age.
- Check for data type format. For example, alpha characters where numbers are expected.
- Check for data patterns such as 16 digits for debit cards, or 10 digits for phone numbers, 9 digits for social security number and so on.
- Foreign key constraint can be applied when the values of a column refer to unique data in another data table. For example, the US states and territories data are a set of defined codes. If data in a State column does not match any of the codes, an error is detected.
- A common data completeness problem is null fields when data must be present for the record to be permitted into the analysis.
- Set the membership constraint. In other words, does a particular value of data correspond to the type of data in this column? The data must belong to a set of data. Similar to the foreign key constraint, consider a field that the data should indicate Male, Female, or Other. Any data that does not belong to this set indicates an invalid data point.
- Data consistency checks are techniques to ensure data in different parts of the data set don't contradict each other. It's important to have a consistent rule to determine which of the two records will be the surviving record. For example, a

customer may have two different addresses. There is a chance that the customer owns two properties which indicates valid data. But, if only one address is to survive, which rule do you apply to delete one and keep the other? Do you save the most recently recorded address or the address that has more reliable source? Or, you may decide to classify the customer in a class of those who own 2 or more properties. Again, the rules of cleansing must be consistent.
- Uniformity constraints check to ensure data scale and units are uniformly recorded. For example, patients' weight may be recorded as kilogram and pound. Height may be recorded in Centimeters or in inches. Data transformation will be needed to create a uniform data unit of measure and scale.
- Other data error issues to monitor are situations where missing data is coded as "999," or "9999." Often you might find "0" when in fact the survey respondent intended "not applicable" or "blank" response. In order to detect data errors you should look for data patterns that have null, "0," "999" or "9999" and combinations.

The statistical techniques include:
- Finding outliers. Identifying data that is outside of ±2sd or ±3sd from the mean.
- Match records using fuzzy logic methods. The fuzzy logic methods offer a score of confidence in how well two records are matched. For example, you can match records by a number of attributes such as customer name, date of birth, customer address, social security number, email address, phone number, and so on. Each matching attribute can raise the level of confidence score, often measured in matching scores between 0% and 100%. You can set automatic matching rules to match records that have 95% or higher matching score. Some databases like Microsoft SQL Server (SSIS) have tools for fuzzy logic record matching and de-duplication.
- Compare the standard deviation and the mean. A standard deviation higher than the mean value should raise eyebrows. This indicates that some extreme values are in the data since the variable cannot be less than zero. You need to establish the minimum and maximum values (range) for your data and identify the data items that are outside of the range. Once these values are corrected or removed, the standard deviation will be smaller.
- Look into minimum prevalence of zero. It's unlikely that certain values can be zero (for example, a subject's height cannot be zero). In some cases, the number of zero values should not exceed a certain percentage of data. You can detect and focus on records that include zeros.
- Look into maximum prevalence of "999," "99.90" or "90.00." The prevalence of values over 90.00 can signal invalid data values. For example, it's highly unlikely that a sample randomly selected from a population would include high blood pressure in 90% of that sample, unless the study targeted the high blood pressure population.

- Look into associations between variables. For example, there should be no association, i.e. no correlation between age and height since these are not correlated in an individual adult. If you find correlation among two variables when there is no association, you should flag the data as suspicious.

6.13 Analysis of Variance (ANOVA) and MANOVA

ANOVA (Analysis of Variance) tests are used for conventional hypothesis testing quite effectively. These tests determine how much the rate of variance between two or more variables match each other. Analysis of Variance (ANOVA) is widely used in statistical inference & design of experiments. It's set up by specifying the null hypothesis. The null hypothesis is that all groups in the analysis are simply random samples of the same population. ANOVA compares variances and mean squares of the variables.

How do you measure rate of variance among categorical variables? You can use rANOVA/rMANOVA for analyzing categorical predictors. MANOVA is the statistical tool for testing multivariate analysis of variance.

ANOVA/MANOVA tests do not handle time-dependent covariates (predictors measured over time). They assume everyone is measured at the same time (time is categorical) and at equally spaced time intervals (a potential problem for most longitudinal studies). Also, you don't get parameter estimates from these tests. You only get the p-values for the test. In the event of missing data, you must impute missing data first. These tests make restrictive assumptions about the correlation structure of data. Consequently, you should be aware of your data limitations and where these tests apply correctly.

The goal of hypothesis testing using statistics tests is to show the result (calculated from the null hypothesis and the sample) is statistically significant if it is deemed unlikely to have occurred by chance (or probability shown by p-value).

A statistically significant result, when a probability (p-value) is less than a threshold (significance level), justifies the rejection of the null hypothesis, but only if the a priori probability of the null hypothesis is not high.

In the typical application of ANOVA, the null hypothesis is that all groups are simply random samples of the same population. For example, when studying the effect of different treatments on similar samples of patients, the null hypothesis would be that all treatments have the same effect (perhaps none). Rejecting the null hypothesis would imply that different treatments result in altered effects.

By construction, hypothesis testing limits the rate of Type I errors (false positives leading to false scientific claims) to a significance level. Experimenters also wish to limit Type II errors (false negatives resulting in missed scientific discoveries). The Type II error rate is a function of several things including sample size (when the sample size is positively correlated with experiment cost), significance level (when the standard of proof is high, the chances of overlooking a discovery are also high) and effect size (when the effect is obvious to the casual observer, Type II error rates are low).

6.14 Advanced Analytics Methods At-a-Glance

Table 6.1 shows the key applications and objectives in advanced analytics and preferred algorithms and methods used by data scientists to achieve each application. We're already studied regression and outlier detection from this list. We'll review some additional topics including Classification, Machine Learning, Time series analysis, Feature selection, clustering and natural language processing. We'll take a deeper look at outlier detection and 2-sample matching & testing algorithms.

Table 6.1: Big data analytics methods overview.

Objective	Statistical Method & Algorithms
Regression	(non-linear) Logistic regression; Kernel regression; Gaussian process regression
Classification	Nearest Neighbor classier; Non parametric Bayes Classifier;Support Vector Machines(SVM)
Machine Learning	Supervised/Unsupervised learning Neural Networks; Decision trees, Bayesian networks, Support Vector Machines(SVM)
Time Series Analysis	Stochastic analysis; Kalman filter; hidden Markov model; Trajectory tracking
Feature Selection, Causality Analysis	LASSO regression,L_1 Norm support vector machines,Gaussian graphical models; discrete graphical models
2-sample Testing & Matching	N-point correlation; bipartite matching;
Clustering	K-means; mean-shift;hierarchical clustering; by dimension reduction; by density estimation
Outlier Detection	By robust L_2 norm estimation; by density estimation; by dimension reduction
Natural Language Processing	Topic modeling, Ontologies, Semantic analysis, Named Entity Relationship(NER), Morphologies, Conference, Sentiment analysis

In summary, modern data analytics methods cover a wide range of topics enabled by artificial intelligence, complexity theory, and systems theory concepts. The methods include regression, classification, machine learning, time series analysis, feature selection, causality analysis, 2-sample testing & matching, clustering, outlier detection and natural language processing. Many of these methods are used in data mining. We'll cover these topics in more detail in the following sections.

6.15 LASSO, L1 and L2 Norm Methods

Both *L1* and *L2* Norm calculations are used in machine learning (ML) methods. *L1*, also known as Robustness measures insensitivity to small deviations from the assumptions and outliers. *L1*-norm is also known as least absolute deviations (LAD), least absolute errors (LAE). It is basically minimizing the sum of the absolute differences *(S)* between the target value (Y_i) and the estimated values $f(x_i)$:

$$S = \sum_{i=1}^{n} |y_i - f(x_i)|.$$

L2-norm is also known as least squares. It is basically minimizing the sum of the square of the differences *(S)* between the target value (Y_i) and the estimated values $f(x_i)$:

$$S = \sum_{i=1}^{n} (y_i - f(x_i))^2$$

The differences of *L1*-norm and *L2*-norm can be promptly summarized as in Table 6.2.

Table 6.2: The differences in two commonly used regression methods.

Least Squares Regression	Least Absolute Deviation Regression
Not very robust	Robust
Stable solution	Unstable solution
Always one solution	Possibly multiple solution
No feature selection	Built-in feature selection
Non-sparse outputs	Sparse outputs
Computationally efficient due to having analytics solutions	Computationally inefficient on non-sparse cases

The difference between *L1* and *L2* can be explained by: *L1* is more robust than the *L2* norm due to its resistance to outlier data and also because in *L2* norm, errors might grow when squared (when error >1). In general, *L1* Norm is more successful to detect outliers than *L2*.

LASSO (Least Absolute Shrinkage and Selection Operator) is a regression method that involves penalizing the absolute size of the regression coefficients.

By penalizing (or equivalently constraining the sum of the absolute values of the estimates) you end up in a situation where some of the parameter estimates may be exactly zero. The larger the penalty applied, the further estimates are shrunk toward zero.

This is convenient when we want some automatic feature/variable selection, or when dealing with highly correlated predictors, where standard regression will usually have regression coefficients that are "too large."

6.16 Kalman Filtering

The idea behind a filter is to extract the required information from a signal, ignoring everything else including the noise in the data. Its applications range from tracking in interactive computer graphics to motion prediction and analyzing noisy sensor data. The Kalman filter is an optimal estimator because it infers parameters of interest from indirect, inaccurate and uncertain observations. It is recursive so that new measurements can be processed as they arrive. It's an optimal estimate since if all noise is Gaussian (i.e. normally distributed), the Kalman filter minimizes the mean square error (MSE) of the estimated parameters. What if the noise is not Gaussian? Given only the mean and standard deviation of noise, the Kalman filter is the best linear estimator. Non-linear estimators may be a better option for non-linear data. Kalman filtering is popular because of the following features:

- Good results in practice due to optimality and structure
- Convenient form for online real time processing
- Easy to formulate and implement given a basic understanding
- Measurement equations need not be inverted

6.17 Trajectory Tracking

Trajectory Tracking is computing future projection of shape and location of data items. It uses many *moving object databases* available from data generated from GPS and smart devices.

6.18 N-point Correlation

N-point correlation functions (NPCF) are powerful spatial statistics capable of fully characterizing any set of multidimensional points. These functions are critical in key data analyses in astronomy, medical and materials science, among other fields, for example to test whether two point sets come from the same distribution and to validate physical models and theories. The n-point correlations have also been used to create feature sets for medical image segmentation and classification.

6.19 Bi-partite Matching

Bipartite matching is a graph method of assigning as many nodes to each other in a graph such that no two edges share a common vertex and having the maximum number of edges. A major goal of many empirical studies in the health and social sciences is to evaluate the causal effect of an intervention, such as a medical treatment or a policy change. The ideal setup for conducting causal inference is a well-executed randomized experiment.

However, random allocation of participants to treatments is often not feasible due to practical or ethical reasons, and observational studies have to be performed instead. For example, in most smoking cessation studies, participants are given the option to choose the treatment they would like to receive. In this case, participants who choose the treatment may differ from those who choose the control condition. Lack of adequate controls for treated participants often leads to biased treatment effect estimation, where the observed effect may be due, in part, to selection effects.

Matching is a popular approach to guard against such selection bias. It classifies participants into homogeneous groups or strata according to certain criteria, such that the individuals within the same matched groups were comparable before they received a treatment. Provided the probability of treatment is influenced only by variables balanced through matching, the difference in their outcomes provides an unbiased estimate of the treatment effect. Matching is appealing for adjusting for measured confounding in observational studies for several reasons. First, well-matched sets provide easily interpretable analyses.

Second, it draws the attention of the users to the covariate balance between matched groups to understand the limits of the analysis. Third, some matched analyses do not need the parametric assumptions required by most regression methods. Fourth, matching that produces nonoverlapped pairs (matching without replacement) maintains an independent structure and thus enables the proper use of the existing statistical inference methods (Hansen and Klopfer 2006). Fifth, the matching process itself does not use the outcome variable information, preventing inappropriate manipulation of the data. Matching-based analysis has been widely applied in various fields, including healthcare research, sociology, economics, business, and political science.

6.20 Mean Shift and K-means Algorithm

K-means and Mean-shift algorithms are commonly used for cluster analysis. A common application is in marketing where companies want to segment their customer base by clustering the data collected from customer behavior. Both methods are used for grouping of data. K-means finds a "center" for each grouping of data by measuring the distance between data points. K is the number of clusters and Means is the average value of the data set in each cluster.

Mean-shift is a sliding-window type of algorithm that attempts to find dense areas of data points. It's a centroid-based approach meaning that its goal is to locate the center points of each group of data. It finds the center of each group by iteratively calculating the mean of the points with in the sliding window. Mean-shift is a non-parametric feature-space analysis technique for locating the maxima of a density function, a so-called mode-seeking algorithm. Other applications include cluster analysis in computer vision and image processing. The mean shift clustering algorithm has two main drawbacks. First, the algorithm is calculation intensive; it requires in general *O(kN**2)* operations, where N is the number of data points and k is the number of average iteration steps for each data point.

Second, the mean shift algorithm relies on sufficiently high data density with a clear gradient to locate the cluster centers. In particular, the mean shift algorithm often fails to find appropriate clusters for so called data outliers, or those data points located between natural clusters.

The k-means algorithm does not have the above two problems. The k-means algorithm normally requires only *O(kN)* operations, so that the k-means algorithm can be applied to a relatively large dataset. However, k-means has two significant limitations. First, the k-means algorithm requires that the number of clusters are pre-determined. In practice, it is often difficult to specify a priori an appropriate cluster number, resulting in some natural clusters being represented by multiple clusters found by the k-means algorithm.

Second, the k-means algorithm is, in general, incapable of identifying non-convex clusters. The second limitation makes the k-means algorithm inadequate for complex non-linear data. These problems can be overcome by simply combining the two algorithms mean shift and k-means together. By combining the two approaches, the data scientist can overcome the accuracy limitations of each model.

6.21 Gaussian Graphical Model

Gaussian graphical model is a simple method for inferring the network of (linear) dependencies among a set of variables is to compute all pairwise correlations and subsequently to draw the corresponding graph (for some specified threshold). While popular and often used in many types of genomic data (e.g. gene expression, metabolite concentrations, etc.) the naive correlation approach does not infer the dependency network.

Instead, graphical Gaussians models (GGMs) should be used. These correctly allow identification of direct influences, have close connections with causal graphical models, are straightforward to interpret, and yet are essentially as easy to compute as naive correlation models.

The key idea behind GGMs is to use *partial correlations* as a measure of independence of any two genes. This makes it straightforward to distinguish direct from indirect interactions. Note that partial correlations are related to the *inverse* of the correlation matrix. Also note that in GGMs, missing edges indicate conditional independence.

6.22 Parametric vs. Non-parametric Methods

Parametric statistical procedures rely on assumptions about the shape of the distribution (i.e., assume a normal distribution) in the underlying population and about the form or parameters (i.e., means and standard deviations) of the assumed distribution. Nonparametric statistical procedures rely on no or few assumptions about the shape or parameters of the population distribution from which the sample was drawn.

Although nonparametric tests have the very desirable property of making fewer assumptions about the distribution of measurements in the population from which we drew our sample, they have two main drawbacks. The first is that they generally are less statistically powerful than the analogous parametric procedures when the data truly are approximately normal. "Less powerful" means that there is a smaller probability that the procedure will tell us that two variables are associated with each other when they in fact truly are associated.

Let's consider a clinical trial example. If you are planning a study and trying to determine how many patients to include, a nonparametric test will require a slightly larger sample size to have the same power as the corresponding parametric test. The second drawback associated with nonparametric tests is that their results are often less easy to interpret than the results of parametric tests.

Many nonparametric tests use rankings of the values in the data rather than using the actual data. Knowing that the difference in mean ranks between two groups is five does not really help our intuitive understanding of the data. On the other hand, knowing that the mean systolic blood pressure of patients taking the new drug was five mmHg lower than the mean systolic blood pressure of patients on the standard treatment is both intuitive and useful.[53]

53 By Tanya Hoskin, a statistician in the Mayo Clinic Department of Health Sciences Research.

Table 6.3 highlights the differences between parametric and non-parametric methods.

Table 6.3: Parametric vs. non-parametric features.

Analysis Type	Example	Parametric Procedure	Nonparametric Procedure
Compare means between two distinct/independent groups	Is the mean systolic blood pressure (at baseline) for patients assigned to placebo different from the mean for patients assigned to the treatment group?	Two-sample t-test	Wilcoxon ranksum test
Compare two quantitative measurements taken from the same individual	Was there a significant change in systolic blood pressure between baseline and the six-month follow-up measurement in the treatment group?	Paired t-test	Wilcoxon signed-rank test
Compare means between three or more distinct/ independent groups	If our experiment had three groups (e.g., placebo, new drug #1, new drug #2), we might want to know whether the mean systolic blood pressure at baseline differed among the three groups?	Analysis of variance (ANOVA)	Kruskal-Wallis test
Estimate the degree of association between two quantitative variables	Is systolic blood pressure associated with the patient's age?	Pearson coefficient of correlation	Spearman's rank correlation

6.23 Non-parametric Bayesian Classifier

A non-parametric Bayes classifier is a Bayesian classifier method that uses non-parametric techniques. In its simplest form, we want to determine the probability that a data item belongs to class A or B. It's used often when the number of clusters are not known in advance. A non-parametric method is essentially a parametric model where the number of parameters increases with data. Also, it's a good choice when we face a really large parametric model, or a model with infinite dimensional function or measure spaces, or when a family of distributions is dense in some large space.

Why use nonparametric models in Bayesian theory of learning? Because it offers a broad class of priors (a prior probability distributions) that can handle any type of distribution, be it normal, or gamma, or other types of distribution. The non-parametric method has shown to perform better than parametric Bayesian classifier on certain types of data. This method essentially allows data to "speak for itself."

6.24 Machine Learning

Machine learning is a subfield of computer science that evolved from the study of pattern recognition and computational learning theory in artificial intelligence. Machine learning explores the study and construction of algorithms that can learn from and make predictions on data.

Data mining is sorting through data to identify patterns and establish relationships. Generally, data mining (sometimes called data or knowledge discovery) is the process of analyzing data from different perspectives and summarizing it into useful information. Data mining is the analysis of data for relationships that have not previously been discovered. It is an interdisciplinary subfield of computer science, the computational process of discovering patterns in large data sets ("big data") involving methods at the intersection of artificial intelligence, machine learning, statistics, and database systems.

There are commercial tools like Skytree[54] that equip a novice data scientist—who may not have years of experience in this topic—with powerful machine learning capability. They offer a more user friendly interface, a variety of algorithms pre-packaged and control over algorithms. There are also open source solutions for more experienced data scientists such as Mahout, Weka, RapidMiner, Python's SciKit-Learn and R. The same tools are often used for general data mining applications.

Neural networks are a main staple set of algorithms used in machine learning, clustering and classification. We'll explore Neural Networks in much more detail in the next chapters.

6.25 Geo-spatial Analysis

Geo-spatial methods overlay data and analytics on geographical maps. A common software tool is ESRI's geographical database as the underlying layer. The company offers tools such as ArcGIS (ESRI reader) to enable delivery and viewing of geo-spatial analysis. Other analytics and visualization software companies are beginning to offer geo-spatial capability. For example, QlikView offers GeoQlik, a QlikView extension.

Geo-spatial analysis uses point data, simply latitude-longitudinal coordinates (Lat-Long for short) associated with data. For example, it's possible to locate the exact location that a customer made a purchase by location data on their phone at the time of purchase. Geospatial analysis has the ability to layer data on top of each

54 http://www.skytree.net/

other layers using latitude-longitude coordinates. This type of analysis is often done in blocks of geography, say by zip code or by a collection of coordinates.

The analysis is often shown as heat maps or color coded maps to depict results and highlight areas of interest. Geo-spatial analysis is done through distance-based analysis and weighted points (min or max distance rules) for a population of people, objects or activity that defines that geography. For example, you can layer cell-phone location data to monitor activity of customers in or around a business area.

Similarly, you may layer customer purchases (say from local pharmacy stores) and their preferences, types of purchases, times of purchase (days of the week and time of day) on top of the household geographic locations to understand the consumer behavior in a given neighborhood. (See Figure 6.6 for an illustration.)

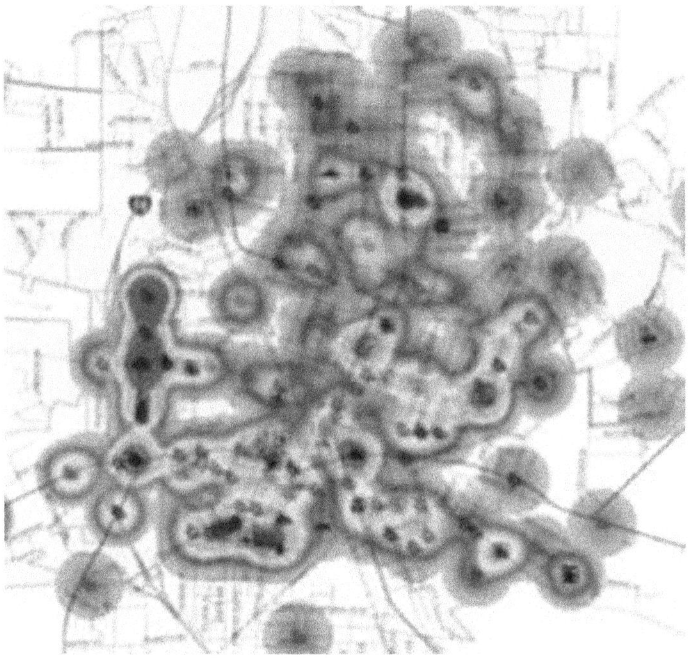

Figure 6.6: Illustration of geo-spatial analysis.

A variation of geo-spatial analysis is geo-temporal analysis. Geo-temporal analysis layers data over time and matches the data to the same coordinates. It allows the user to visually analyze trends, changes in population, wealth, adoption of social trends or disease progression. For example, geo-temporal analysis was used to track number of deaths due to asbestos in Texas in the U.S.

In another study, geo-temporal analysis was utilized to track trends showing abnormally high chronic obstructive pulmonary disease (COPD) in a population located downstream from a factory that caused pollution. The geo-temporal maps showed correlation between the factory polluting the air with sulfides (environmental pollution & burden of disease) and higher than normal rates of COPD in the downstream population.

6.26 Logistic Regression or Logit

Logistic regression has the same structure and concept as linear regression but with the ability to handle categorical variables. It's often referred to as Logit. Logistic regression looks to predict a binary response to one or more predictor (independent) variables. It uses a probabilistic classification model, where the probabilities describe possible outcomes of each trial modeled as a logistic function.

Logistic regression comes in three flavors: binary, multinomial or ordinal logistic regression.

- Binary logistic regression assigns a binary value (1,0) to the dependent variable
- The multinomial version has multiple outcomes for its dependent variable
- The ordinal version deals with dependent variable that has ordinal value

Logistic regression is commonly used in bio-sciences and clinical data analysis. For example the Trauma & Injury Severity Score (TRISS) is used to predict mortality in injured patients based on logistic regression. Also, Logit analysis is often used to predict whether a patient has a given disease based on certain demographic data.

6.27 Predictive Modeling Approaches

Data scientists commonly use any of nine different data analytics methods for predictive modeling. Here is the list of predictive modeling methods. We'll review these in more detail in the upcoming sections.

- Bayesian Methods
- Classification (K-nearest neighbor algorithm)
- MARS (Multivariate Adaptive Regression Splines)
- Neural networks (Machine Learning)
- ACE (Alternate Conditional Expectation) & AVAS (Additivity & Variance Stabilization)
- CART (Classification & Regression Trees)
- Boosted trees
- Random forests
- Support Vector Machines

6.28 Alternate Conditional Expectation (ACE)

The alternate conditional expectation (ACE) algorithm is a non-parametric automatic transformation method that produces the maximum multiple correlation of a response and a set of predictor variables. The approach solves the general problem of establishing the linearity assumption required in regression analysis, so that the relationship between response and independent variables can be best described and the existence of non-linear relationships can be explored and uncovered. An examination of these results can give the data analyst insight into the relationships between these variables, and suggest if transformations are required.

6.29 Clustering vs. Classification

Clustering and classification are both used in data mining. Both split data into groups. But, there is a huge difference between the two methods. Clustering is unsupervised learning. When you're given new data and your task is to split data into like groups based on similarities between data items, you use clustering. Classification is supervised learning. When you already have a trained model and are looking to classify a new subject you use classification. Table 6.4 and Figure 6.7 show the differences in more detail.

Table 6.4: Distinctive features of clustering and classification.

Classification	Clustering
– We have a training data set that has been previously categorized. – Based on this training set the algorithm finds the proper category for a new data to belong to	– We don't know the similarity or characteristics of data in advance – The algorithm splits the dataset into subsets such that subsets have similar data
– Known as supervised learning	– Known as unsupervised learning
– Example: We have profiles of customers who buy a product. Does this new customer fit into that profile?	– Example: We have data set of customers. Segment customers by their characteristics.

While both methods group data by one or more features (data characteristics) into like chunks (or clouds), they use different methods for achieving their objective. Let's examine each more closely:

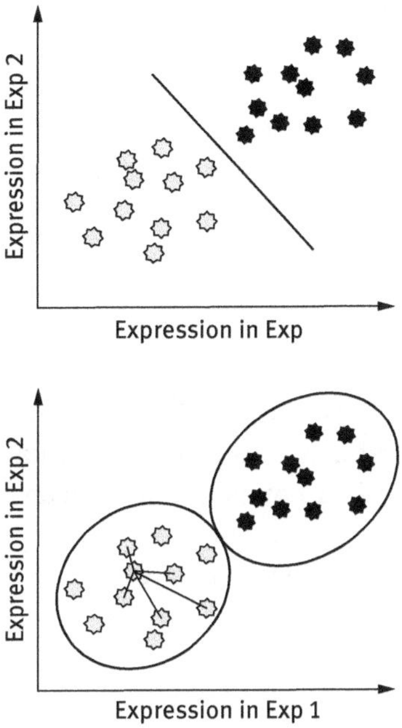

Figure 6.7: Graphical representation of classification (top quadrant) and clustering (bottom quadrant).

- **Classification**
  - Have labels (tags) for some data
  - Goal: want a "rule" that accurately assigns labels to new data
  - Common algorithm: Neural networks

- **Clustering**
  - Data has no labels
  - Goal: Group points into clusters based on how "near" they are to one another
  - Identifies structure in data
  - Common algorithm: K-Means clustering

6.30 K-means Clustering Method

K-means clustering method generates k clusters from data. There are two steps to this algorithm.

Given an initial set of k means $m_1, \ldots, m_k$, the algorithm alternates between two steps:

Step #1: Assign each observation to the cluster whose mean produces the least within-cluster sum of squares (using least squares estimator).

Step#2: Update the mean with a new calculation to be the centroid of the observations in the new clusters.

The algorithm stops when the assignments no longer change. At that point, the algorithm has converged. The algorithm works by assigning objects to the nearest cluster by distance.

6.31 Classification Using Neural Networks

You can cluster and classify data using artificial neural networks (ANN). The following example is a graphical representation of how neural networks can cluster data. Consider a two-dimensional space with some existing data points that are already clustered into two groups.

The objective of clustering is to find the best "line" that separates the two groups as shown in the diagram by:

$$Y = W_1X_1 + W_2X_2 + W_0$$

When this line equals zero, the line separates the data into two clusters perfectly (see Figure 6.8). Now, let's suppose new data arrives and we need to classify it. Which cluster does the new data belong to? The answer: "it depends on which side of the

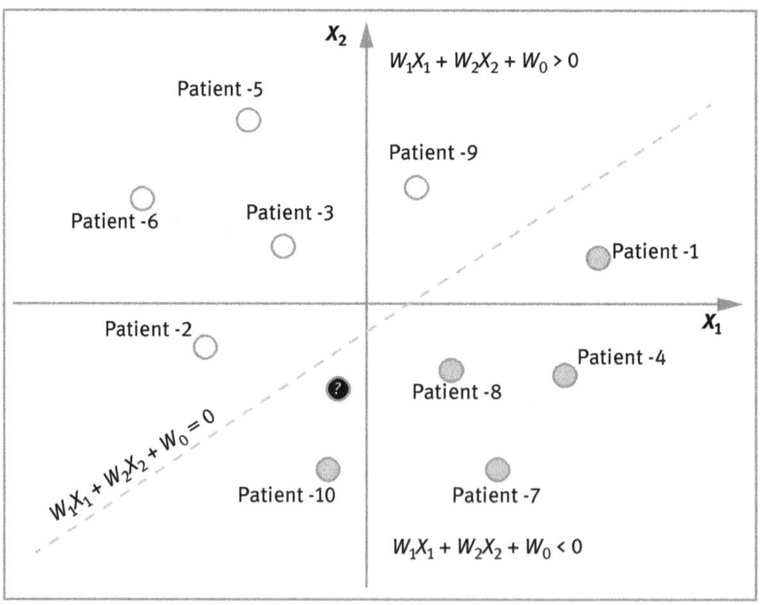

Figure 6.8: Example of classification using a simple clustering line.

line it ends up." The challenge is to find the coefficients, W_1, W_2, W_O in the equation for the line that best separate the two data clusters. We'll see that the equation for the line is achieved by training a neural network that optimizes the values for these coefficients.

We'll review neural networks in more detail in the next chapters.

6.32 Principal Component Analysis

Principal component analysis (PCA) is mostly used as a tool in exploratory data analysis and for making predictive models. It's a form of *factor analysis*.

Principal component analysis is highly regarded as an accurate tool for unsupervised learning. In general, unsupervised learning is much more challenging than supervised learning as it is more subjective. There is no simple goal for the analysis, such as predicting the response. Unsupervised learning is often used as part of an *exploratory data analysis*. Further, it is harder to assess the accuracy of the results obtained, since there is no universally accepted method for cross-validation or validation. Simply put, we cannot really *check out work* in an unsupervised setting, beyond simple intuition or theoretical knowledge of the process at hand.

There are many uses for unsupervised methods as illustrated by several use-cases. PCA helps us understand cancer behavior by identifying subgroups of patients. Websites (particularly e-commerce) often try to recommend product to you based on your previous activity using PCA algorithms. Another example is Netflix movie recommendations.

When presented with a large set of correlated variables, principal components allow us to summarize the set into a smaller number of representative variables that *collectively* explain most of the variability in the original set.

Principal component analysis (PCA) refers to the process for which principal components are computed, and the subsequent use of the components in understanding the data. PCA also serves as a tool for data visualization.

Here is how PCA works. Suppose we wish to visualize n observations with measurements on a set of p features for part of an exploratory data analysis. We could use scatterplots of the data in two-dimensions. However, there are potentially *a lot* of scatterplots you would have to create. If p is large; then $(p(p-1)/2)$ scatterplots would be required (to be exact).

For example, if $p = 10$ there are 45 plots. Clearly we need an alternative when p is large. Specifically, we want to find a low-dimensional representation of the data that captures as much information as possible. PCA provides a method to do this. PCA seeks a small number of dimensions that are as interesting as possible, where the concept of *interesting* is measured by the amount that the observations vary across the dimension. Each of the dimensions found by PCA is a linear combination of the p features, so this is technically not a form of feature selection. Also note that before computing PCA the data must be centered and have a mean zero (unless all of the data is of the same unit).

We can also measure how much information is lost by utilizing principal components. To do this we can compute the *proportion of variance explained* (PVE) by each principal component. It is generally best explained as a cumulative plot, such that we can visualize the PVE for each component and for the total variance explained. Once we have this measurement, we can start to conclude if the principal components explain enough data to provide an accurate summary.

In summary, principal component analysis uses transformation of highly linearly correlated data into highly un-correlated data, called *components*. Each component is computed to explain the largest amount of variance in the data (the first component). The next highest amount of variance is assigned to the 2nd component and so on.

Each subsequent component is selected to be orthogonal to the previous one. PCA can be used to reduce dimensionality of data (Example: extracting features in image pattern facial recognition). PCA uses variance and co-variance of data in order to calculate the components of data. One fact to note is that in PCA, there is an equal or smaller number of components than there are original variables.

6.33 Stratification Method

Stratification is a technique for classifying data (such as population, patients, consumers, etc.) into subgroups. Stratification is a simple method for classification of data into categories and sub-categories by certain criteria, like age, gender, lifestyle, race, etc. Stratification is important when we consider Simpson's paradox. Simpson's Paradox states that data may show one thing in aggregate form and something totally different when disaggregated.

Stratification helps detect systematic errors in our studies. There are a dozen sources of bias in data. But here are four types of bias associated with systematic error:

- Information bias: This occurs when we have mis-measurement or misclassification of study factors
- Selection bias: Selection bias occurs when selecting non-representative samples from populations
- Confounding: This bias is common due to mixing and distortion of association between factors and outcomes
- Interaction: Interaction occurs due to the independent operation of two or more factors that interact to produce an unanticipated effect

What are some of the strategies and recommendations to overcome these biases? Since confounding is a systematic (non-random) error, hypothesis testing can't be used to detect it. Here are a couple of actions to consider:

- Identify confounder variables before data is collected
- Adjustments for confounding are contraindicated when interaction is present

Another technique to reduce confounding bias is to apply *propensity matching* techniques.

6.34 Propensity Score Matching Approach

Propensity score matching (PSM) is a technique used to reduce the effects of confounding in observational studies. When we compare the effects of the treatment in a population, we want to be certain that the response to the treatment is based on the treatment, and not due to our selection bias. Hence we use PSM to determine the selection bias between two groups that respond positively and negatively to a treatment. It allows one to mimic some of the characteristic of randomized controlled trial studies. PSM works by matching treated and untreated observations on the estimated probability of being treated (propensity score).[55]

PSM is implemented by matching one-to-one or pair matching, in which pairs of treated and untreated subjects are formed, such that matched subjects have similar values of the propensity score. Said differently, a propensity score is the probability that a person being assigned to a particular group given a set of observed covariates. Propensity scores are used to reduce selection bias by equating groups based on these covariates.

PSM estimates the effect of a treatment or drug by accounting for co-variables that predict receiving the treatment. This approach reduces bias due to confounding variables which can be found in the estimate of the treatment. Use PSM to account for systematic differences in baseline characteristics between treated and untreated subjects when estimating the effect of treatment on outcomes. The PSM method is available in several statistical packages including R as part of the MatchIt package.

Propensity score matching entails forming matched sets of treated and untreated subjects who share a similar value of the propensity score (Rosenbaum & Rubin 1983). Propensity score matching allows one to estimate the *average treatment effect of the treated* (ATT).

The most common implementation of propensity score matching is one-to-one or pair matching, in which pairs of treated and untreated subjects are formed, such that matched subjects have similar values of the propensity score. Although one-to-one matching appears to be the most common approach to propensity score

55 This is the most commonly used PSM technique. We match observations based on probability of participation.

matching, other approaches can be used. These are discussed at the end of this section. Unless stated otherwise, the following discussion is in the context of 1:1 matching.

Once a matched sample has been formed, the treatment effect can be estimated by directly comparing outcomes between treated and untreated subjects in the matched sample.

Steps used in implementing PSM are as follows:

1. You need representative and comparable data for both treatment and comparison groups. You can run logistic regression if your data includes categorical data. To run the regression model, assume the dependent variable: *Y=1* for participate, or *Y=0* otherwise (not participate).
2. Use a logit (or other discrete choice model) to estimate program participations as a function of observable characteristics.
3. Use predicted values from logit to generate a propensity score $p(x_i)$ for all treatment and comparison group members.
4. Match Pairs: Individuals with similar covariates are matched together from both groups (from treatment group and comparison group).
5. You need to determine a *tolerance limit*: how different can control individuals or subjects be and still be a match? You can choose any method like the nearest neighbor to match individuals from both groups.
6. Once matches are made, we can calculate impact by comparing the means of outcomes across participants and their matched pairs.

The ideal comparison group is selected such that it matches the treatment group using either a comprehensive baseline survey or time invariant characteristics. The matches are selected on the basis of similarities in observed characteristics. Of course, this approach assumes no selection bias based on unobserved characteristics of subjects in the study.

In summary, for matching pairs, compare probability of data against what is predicted by regression. You can run logistic regression if your data includes categorical data. To run the regression model, assume the dependent variable: *Y=1* for participate, or *Y=0* otherwise (not participate). Then choose appropriate confounders (variables associated with treatment and outcome).

Next compute the propensity score: predicted probability *(p)* or *log[p/(1-p)]*. As the next step, check that the propensity score is balanced across treatment and comparison groups. Check that covariates are balanced across treatment and comparison groups using standardized differences to examine distributions. Next, match each participant to one or more non-participants in propensity score (using nearest neighbor matching or stratification matching or other methods). You must verify that covariates are balanced across treatment and comparison groups in the matched sample. Finally, you can apply multivariate analysis based on the new data sample.

6.35 Adherence Analysis Method

The *adherence analysis method* is often used in healthcare to measure patient adherence to a given practice or medication regimen. It measures the degree to which patients follow the treatment and recommendation of their health professionals. This method is used across many disease conditions. Adherence analysis studies have shown that patient in-adherence average is about 50%.[56] Similarly, you can apply this technique in other applications. For example, you can measure the degree that students follow instructions they learned from a course to measure the efficacy of that training program.

Adherence can be measured across time: longitudinal, cross-sectional and retrospectively. Adherence may be measured by regimen: medication, health-behavior, appointment-keeping and similar regimen. Adherence analysis performs correlation analysis between treatment and patient outcomes. It calculates the "*r* effect size" in the data and can identify what factors are correlated to adherence. For example, an adherence analysis shows that patient adherence is highly correlated to physician's communication skills.

6.36 Meta-analysis Methods

Meta-analysis is the systemic examination of multiple studies. This form of analysis typically combines results from multiple research articles about the same type of data and the same hypothesis. For example, let's assume that you're performing meta-analysis on the impact of electric cars on local economy. Some research might claim that electric cars have a positive impact while some might conclude the opposite. Both have applied rigorous scientific and statistical methods and data analytics to reach different conclusions. So, how do you reconcile these articles? This is where meta-analysis comes in.

To perform meta-analysis, data from these multiple studies are compiled from prior publications, research papers and data bases. Each article may have a different statistical significance and effect size that must be considered. There are certain Inclusion-Exclusion criteria applied in this type of analysis. Common Inclusion-Exclusion criteria include:

- Studies that are peer-reviewed are preferred. You may decide to include only these articles in your meta-analysis
- Time horizon of studies. You may choose articles that are more recent, say in the last 5 years and exclude any research result older than 5 years.

---

56 Jennifer Kim, Kelsy Combs, Jonathan Downs et. al., "Medication Adherence: The Elephant in the Room," *US Pharmacist*, Vol. 43, No. 1, 30–34, 2018.

- Population demographics: Geographic and language. You may include all populations globally, or limit your meta-analysis to certain populations or languages. The strength and the power of your meta-analysis depends on these choices.

For meta-analysis, you must consider coding the articles and determine the effect size of each prior study in these articles. This is known as article coding and effect size extraction. The article coding and effect size extraction applies the following steps:

First articles are coded for purpose and frequency of variables. This is an important step to maintain consistent strength and quality of data for analysis. Next, *Effect size r* is either extracted or computed from article data. It's recommended that you keep significance $p<0.05$. For non-significant results, you can allow $r=0$. Next a random effects model, using un-weighted mean r, is used based on k number of studies. A fixed effects model uses the weighted mean r. Weights are determined by the sample size of each study.

6.37 Stochastic Models—Markov Chain Analysis

In healthcare and population health studies, many diseases are interpreted via stages of progression of the disease. A Markov chain is a random process that transitions from one state to another state independent of prior states. Markov chains can estimate rates of transition between stages as the subjects move from one stage to another over time. They measure and study the probability of transition between stages as shown in the diagram below. For example, compare population of patients' disease progression who participate in a treatment vs. control population.

As Figure 6.9 illustrates, there are n stages possible. The state of a patient may transition from one state to another. Any subject can transition to the next stage or to a final stage n.

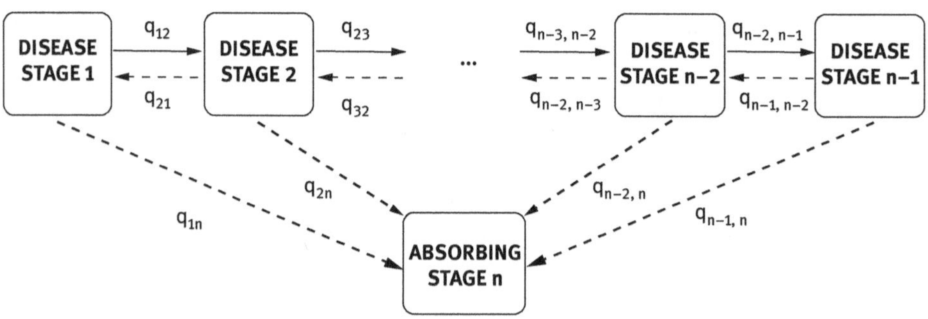

Figure 6.9: Markov chain with n stages.

6.38 Handling Noisy Data—Kalman Filters

Kalman filters offer stochastic algorithms that infer parameters of interest from indirect, inaccurate and uncertain observations. It uses a series of measurements observed over time to produce estimates of unknown variables in particular from noisy data.

Kalman filters are used in time series analysis and online real time processing. They use predictors, filters and smoothing algorithms in a 3-step recursive process:

Step 1: A predictor computes estimates of the current state variables along with their uncertainties using previous measurements.

Step 2: A filter ***updates*** estimates system parameters using previous and current measurements using a weighted average (with more weight given to estimates with higher certainty).

Step 3: A ***smoothing*** algorithm computes estimates of the system parameter's current values using previous, current and future measurements.

The weights are calculated from the covariance (as a measure of estimated uncertainty) of the prediction of the system's state. Kalman filters are frequently used when a predictor with good noise filtering capability are needed.

A Kalman filter is an optimal estimator—i.e. infers parameters of interest from indirect, inaccurate and uncertain observations. It is recursive so that new measurements can be processed as they arrive. But what does optimal mean? It implies that if all noise is Gaussian, the Kalman filter minimizes the mean square error of the estimated parameters. What if the noise is NOT Gaussian?

Given only the mean and standard deviation of noise, the Kalman filter is the best linear estimator. Why is Kalman Filtering so popular? There are four reasons:

- They produce good results in practice due to optimality and structure.
- They're in a convenient form for online real time processing.
- They're easy to formulate and implement given a basic understanding.
- The measurement equations need not be inverted.

6.39 Tree-based Analysis

Tree based analysis is often used for regression analysis and classification problems. It's important to note that classification trees are different from k-nearest neighbor methods. The goal of tree-based analysis is to partition the space with the intent to identify representative centroids. This method partitions the space and recursively divides it into smaller regions until every region is assigned a class label.

In theory, given a data vector $X: \{x_1, x_2, \ldots, X_n\}$, Tree-based analysis will split the vector by thresholds on the values. When the thresholds are applied recursively, the tree continues to grow from there.

Figure 6.10 illustrates a *decision-tree* example. In one example, there x1,. . ., x6 data values and we partition the data by applying the threshold to split them into 2, then into 6 classes.

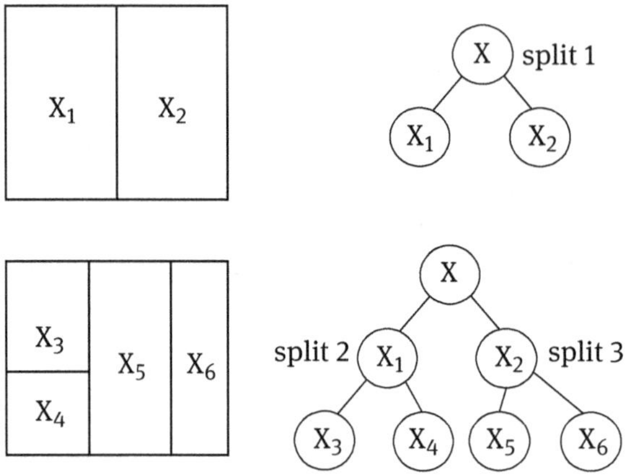

Figure 6.10: Classification using a decision-tree.

In the next example, Figure 6.11, we classify patients based on the value of their systolic blood pressure and tachycardia. The first split occurs if the patient's systolic blood pressure is > 91. The next split is if the patient's age is >62.5. The final split is based on whether sinus tachycardia is present. The decision-tree classifies patients into high-risk or low-risk categories.

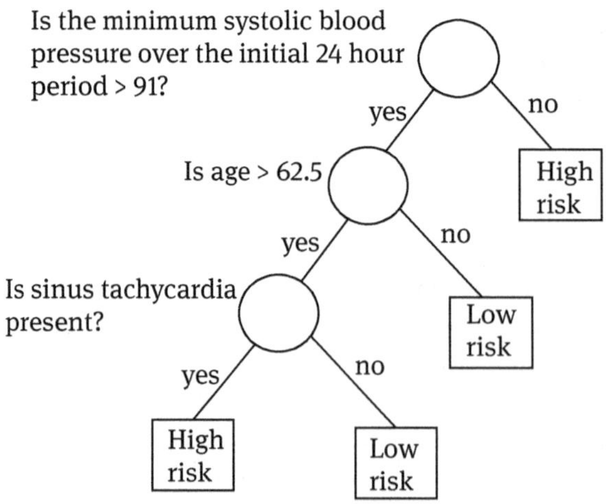

Figure 6.11: Classification using decision-tree method.

6.40 Random Forest Techniques

Random forest is a machine learning method used for classification and regression. The random forest algorithm builds multiple decision trees during training and outputs the class that is the mode of the classes (classification) or mean prediction (regression) of individual trees. As the name sounds, random forest is a collection of multiple trees.

The way random forest works is that multiple decision trees are built to explain the different features in data. The decision trees get trained on data and can make predictions on new data. Random forest is used to rank the importance of variables in a classification. Classification trees are grown to classify each data variable as we discussed in the previous section under tree-based methods (this is known as learning).

When formed, each tree gives a score for that classification. Tree branches are grown similar to K-nearest neighbor (*kNN*) method to explain the difference in features.

Random forest techniques have many advantages. They can be effective on large data and thousands of input variables. They have the robust feature to estimate missing data. But compared to other new techniques, they're not the most accurate method.

Random forest runtimes are quite fast, and they are able to deal with unbalanced and missing data. Random forest's weaknesses are that when used for regression they cannot predict beyond the range in the training data, and that they may over-fit data sets that are particularly noisy.

The random forest method is an ensemble approach that can also be thought of as a form of nearest neighbor predictor. Ensembles are divide-and-conquer approaches used to improve performance. The main principle behind ensemble methods is that a group of "weak learners" can come together to form a "strong learner." Each classifier, individually, is a "weak learner," while all the classifiers taken together are a "strong learner."

In Figure 6.12, three decision trees are shown to form a simple random forest. Let's assume that we want to classify consumers based on the amount of their purchases, location and other attributes. These are called decision trees because the classification follows several branches of "if … then…" decision splits. Imagine that we want to start with amount of purchase as the first decision or split. The split can be thought of as a feature in machine learning. Let's say "if the purchase amount is over $1,000.00 we continue on the right branch, and then repeat the decision process for other features until there are no more decisions before us".

At each branch, the feature threshold that best split the remaining samples locally are found. Random forest then combines results from many individual decision trees. Since RF combines multiple models, it falls under the category of ensemble learning.

Random forest uses decision trees, but takes a different approach. Rather than growing a single, very deep tree that is carefully overseen by an analyst, random

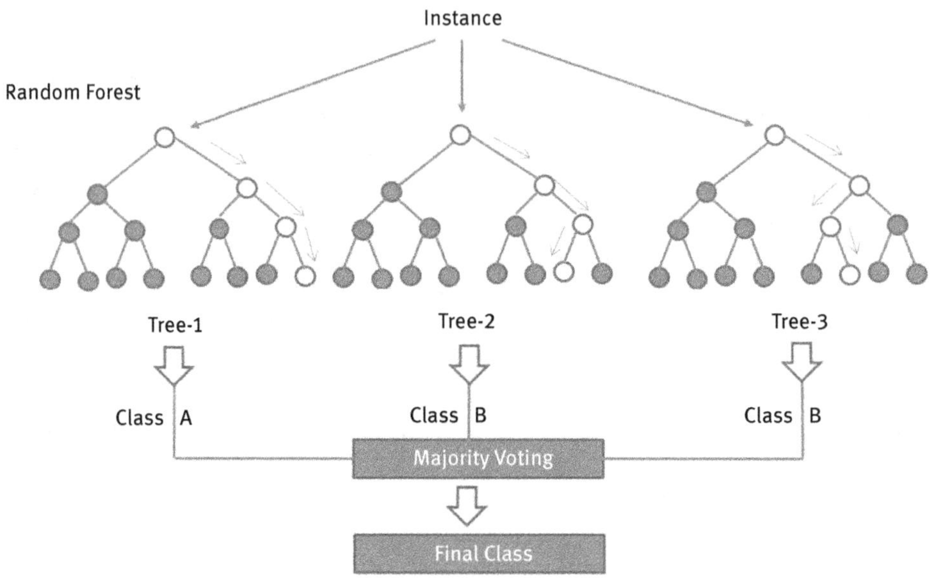

Figure 6.12: Random Forest Classification.

forest relies on aggregating the output from many "shallow" trees (sometimes called "stumps") that are tuned and pruned without much (or any) analyst oversight. Some of these trees may have been grown from samples that said *age* was the more important feature (as opposed to *income*).

There are two main methods for combining multiple decision trees into a random forest:

1. Bagging: This is also called bootstrap aggregation. In the bagging method, decision trees are trained on sampled subsets of the data. This method is efficient and acceptable as long as the individual trees are not correlated.
2. Boosting: This uses gradient boosting methods. In boosting methods, samples are weighted such that samples that were incorrectly classified get a higher weight and therefore sampled more often.

The final result of the model is calculated by averaging over all classifications from these sampled trees or by majority vote.

Classification and Regression Tree

Classification and regression tree (CART) is a set of techniques for classification and prediction. The purpose of the analyses via tree-building algorithms is to determine a set of *if-then* logical (split) conditions that permit accurate classification of cases.

When applied to classification, given a variable (for example, Cost or Preference), the algorithm will identify groupings that best differentiate between high and low cases, using any variable in the dataset. The next section explains how classification by decision tree is performed.

Decision Tree Construction

Let's take a closer look at decision tree construction with an example. Constructing decision trees is a top-down divide-and-conquer technique. Consider the data set that includes data marked as α's and β's as illustrated in Figure 6.13. Our goal is to define a rule that separates the two groups of data. We have two rules:

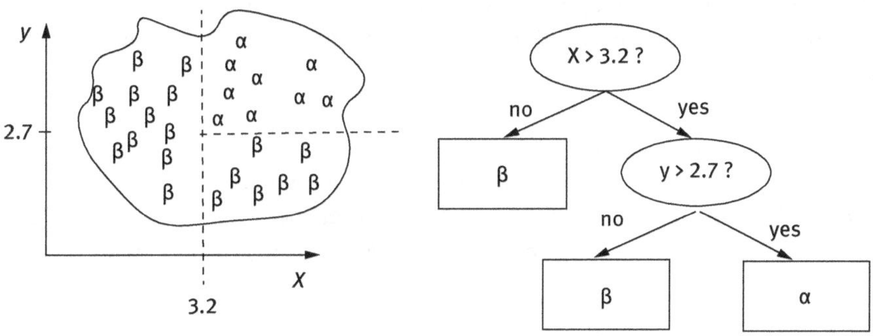

Figure 6.13: Constructing a decision tree to classify two data groups.

- First rule: if x >3.2, we can separate β's (on the left) from α's into two classes.
- Second rule: if y >2.7, then we have two new classes separating α's from β's.

These rules define a decision tree. A top-down divide-and-conquer technique is used in this example. Given the values as shown in the 2-dimensional space, we want to define a rule that separates the two groups of data. We apply a decision tree algorithm called the covering algorithm.

We cover all individual data items to define the rule. We cover the α's and find that if x>3.2, we can separate α's from β's into two classes. But, in the right hand side, there are still some β's mixed with α's. So, we develop another rule that if y>2.7, then we have two new classes. By creating these rules, we've in essence developed a decision tree as shown.

As we discussed earlier, Random forest is formed from multiple decision trees. Together the ensemble of trees provides classification of data.

The random forest scheme was proposed by Leo Breiman in 2001. Random forest consists of forming multiple decision trees on sub-sets of data. Each tree is constructed from a random selection of input variables (this is called boosting). In some variations of the model, the decision trees which deliver the greatest amount of information (highest information gain), i.e. best separation ability are selected for final classifiers of data and work. The final decision trees are used in an ensemble to provide data classification (this is called bagging).

Each decision tree is built from a separate random sample of data, often using bootstrap sampling. You can build hundreds of trees each from a separate bootstrap sample of data. Through this process we gain diversity from examining different subsets of data.

For example, suppose you are trying to analyze customer churn. Marketing experts will tell you that customers are not exactly loyal or disloyal because there is a spectrum of loyalty. But, to oversimplify, let's assume we want to classify individuals as loyal (likely to remain customers) or disloyal (likely to leave). Suppose further that 10% of individuals in general are "disloyal" and 90% are "loyal." Then, any (large enough) random sample will have both "loyal" and "disloyal" individuals in these proportions.

Using decision trees you may be able to identify segments of individuals with a higher concentration of a particular class. We'll try to identify "loyal" individuals in this example. Suppose that you had additional information about each individual... say *income*. A decision tree algorithm may identify that the segment of individuals with less than $50,000 of income is composed of 30% "disloyal" and 70% "loyal." That is, people in this segment are 3X more likely to be 'disloyal' vs. the general population.

Now, suppose that you also have information about each individual's age. The decision tree algorithm might also identify that, within the "less than $50,000 of income" group, that the segment of individuals under the age of 30 is composed of 50% "disloyal" and 50% "loyal." This more-refined segment (income less than $50,000 AND age less than 30) is 5x more likely to be 'disloyal' when compared to the general population. ...And so on... decision trees may identify many, many such features that are predictive of whether an individual will be a "loyal" or "disloyal" customer.

The decision tree algorithm can determine both the order of the branching as well as the bins on which to break (income > 50k, age < 30, ... etc.). But generally, tuning a decision tree requires a trained analyst to make sure the breaks are meaningful. Also the analyst needs to help "prune" the tree to avoid over-fitting and/or nodes with inappropriate coverage.

A random forest uses decision trees, but takes a different approach. Rather than growing a single, very deep tree that is carefully overseen by an analyst, a random forest relies on aggregating the output from many "shallow" trees (sometimes called "stumps") that are tuned and pruned without much (or any) analyst oversight. Some of these trees may have been grown from samples that said *age* was the more important feature (as opposed to *income*).

Other trees may find that a more relevant bin of income is $100,000 (as opposed to $50,000). Other trees may find completely different features to be relevant. The idea is that the errors from many "shallow" trees will wash out when aggregated and lead to a more accurate prediction.

Rather than using deep decision trees, random forest methods use many shallow trees that classify for a few single rules. Each tree is a poor classifier, but together the forest provides an ensemble that's more accurate.

Random forest methods are ideal classifiers for noisy and sparse data as well as for unsupervised learning. Because a random forest builds many trees from a subset of the input variables and their values, it inherently contains some underlying decision trees that omit the noise generating variables (or features). The result is that the noise cancels out and the final classification is more accurate.

You can use Python's Scikit and Numpy components to access its random forest classifier functions. A common function is the RandomForestRegressor class.

6.41 Hierarchical Clustering Analysis (HCA) Method

Hierarchical clustering analysis (HCA) is a data mining technique used for cluster analysis. It works by building a hierarchy of clusters using one of two methods: a bottom up or a top-down approach. In the bottom-up approach each observation starts in its own cluster, and pairs of clusters are merged as one moves up the hierarchy. In the top-down approach, all observations start in one cluster, and get split recursively as one moves down the hierarchy.

Then data "splits" or combinations of observations are determined by a measure of dissimilarity between sets of observations.

The metric for dissimilarity measure can be defined by one of the following techniques, such as Euclidean distance, squared Euclidean distance, Manhattan distance, maximum distance or Mahalanobis distance. The equations are shown below:

Euclidean distance:

$$||a-b||_2 = \sqrt{\sum_i (a_i - b_i)^2}$$

Squared Euclidean distance:

$$||a-b||_2^2 = \sum_i (a_i - b_i)^2$$

Manhattan distance:

$$||a-b||_1 = \sum_i |a_i - b_i|$$

Maximum distance:

$$||a - b||_\alpha = max\ |a_i - b_i|\ for\ all\ i$$

Mahalanobis distance:

$$\sqrt{(a-b)^T S^{-1}(a-b)}$$

HCA is a powerful data mining technique that works by forming hierarchies of data. Data is increasingly becoming multi-dimensional. It's common to find data sets that have more then 10, 100 or even thousands of dimensions. We can plot multi-dimensional data graphically, but the graph gets cluttered with higher dimensional data. As a result, we can apply dimension reduction techniques.

Distance (or dissimilarity) between multiple data items can be measured through many techniques. Splits in data are determined by the measure of dissimilarity. In this approach, data items that are similar to each other (closer in distance) are clustered together versus the data items that are dissimilar. The bigger the dissimilarity, the bigger the distance between the two data items. There are different algorithms for hierarchical clustering analysis. One is called *agglomerative clustering*. It begins with placing each individual data item in its own cluster. Then group each data item to another data item with the closest proximity. Repeat this process until more data items are grouped together as you allow bigger distances between the groups. In other words, the distance increases as more groups merge together.

Figure 6.14 illustrates a graphical representation of the HCA clustering method.

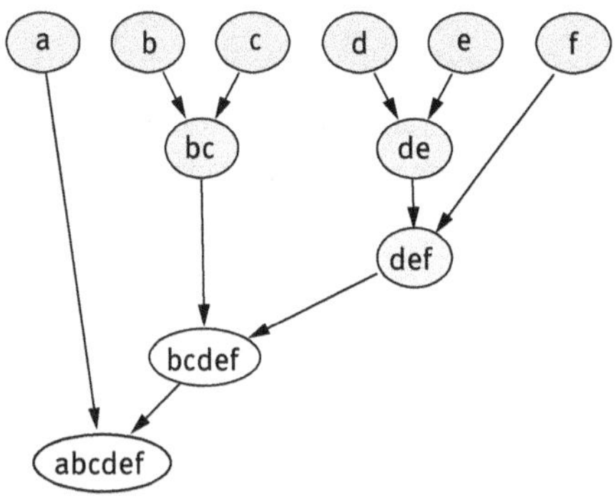

Figure 6.14: HCA Clustering Technique.

The other technique uses a Dendrogram approach. In the previous approach, we used the similarity between two data points to identify whether they fit into the same cluster. In Dendrogram approach, we determine if two data items belong to the same cluster by their dissimilarity, namely by their distance. In this approach the level of dissimilarity (negative similarity) is used to form splits in data sets. The result is an interpretive visual tree diagram of the hierarchy, described in more detail below:

- Step 1: Dimension hierarchy generation: First, all the original dimensions of a multidimensional data set are organized into a hierarchical dimension cluster tree according to similarities among the dimensions. Each original dimension is mapped to a leaf node in this tree. Similar dimensions are placed together and form a cluster, and similar clusters in turn compose higher level clusters. Users have the option of using the system provided automatic clustering approach, using their customized clustering approaches, or specifying a hierarchical dimension cluster tree manually.
- Step 2: Dimension hierarchy navigation and modification: Next, users can navigate through the hierarchical dimension cluster tree in order to gain a better understanding of it. Users can also interactively modify the hierarchy structure and supply meaningful names to the clusters. The hierarchical dimension cluster tree is visualized in a radial space-filling display named InterRing, which contains a suite of navigation and modification tools.
- Step 3: Dimension cluster selection: Next, users interactively select interesting dimension clusters from the hierarchy in order to construct a lower dimensional subspace. Several selection mechanisms are provided in InterRing to facilitate dimension cluster selection. Selected clusters are highlighted.
- Step 4: Representative dimension generation: In this step, a representative dimension (RD) is assigned or created for each selected dimension cluster. The selected dimension clusters construct the lower dimensional space through these RDs. RDs are selected to best reflect the aggregate characteristics of their associated clusters. For example, an RD can be the average of all the original dimensions in the cluster, or can be an original dimension located in the center of the cluster. Users have the option to select one of the system-provided RD generation methods or use a customized one.
- Step 5: Data projection and visualization: Finally, the data set is projected from the original high dimensional space to a lower dimensional space (LD space) composed of the RDs of the selected clusters. We call its projection in the LD space the mapped data set. The mapped data set can be viewed as an ordinary data set in the LD space and can be readily visualized using existing multidimensional visualization techniques.

This is an advantage of HCA; it is so flexible that it can be applied to any existing multidimensional data visualization technique. In order to provide further dimension cluster characteristics in the LD space, such as the dissimilarity information between dimensions within a cluster, we attach the dimension cluster characteristics information to the mapped data set and provide the option to display it using extensions to the data visualization techniques.

6.42 Outlier Detection by Robust Estimation Method

We've discussed some methods for outlier detection already. We saw that a statistics technique such as designating any data outside of ±3 sigma (or standard deviations) as outliers can be practical. Let's consider the robust methods as another technique.

Conventional least squares (LS) estimators often miss or smooth outliers in data. However, Robust (L1) estimation methods overcome the classical parametric estimation limitations. Robust estimation methods are used to detect outliers and anomalies in data.

The L1 estimation method iteratively detects possible outlier data by analyzing globally fitted model in four steps:

1. Compute an initial fit to the whole set of data by using the least squares (LS) method
2. Determine the residuals for each datum
3. Remove data whose residuals are greater than a threshold; stop if no data should be removed
4. Compute a new fit to the remaining data, and go to Step 2

Another technique is to apply different maximum likelihood estimators (MLE), also known as M-estimator to detect outliers. There are other techniques to consider. One technique is to apply a min-max solution using an estimator that maximizes asymptotic variance over some family of densities.

Other estimators are well known L_1, L_2, $L_{1\text{-}2}$, L_p and Cauchy methods for detecting outliers. Finally, two other estimators that are popular are L-estimators (derived from linear combination of order statistics) and R-estimates (derived from rank tests). The details of these methods are outside the scope of this book but most statistical packages offer tools to compute outliers using any of the methods described here.

6.43 Feature Selection Techniques

A feature selection procedure is also known as *variable subset selection.* Feature selection attempts to identify the relevant and important variables and predictors from irrelevant or redundant variables. Feature selection is an important tool to machine learning applications.

It helps to reduce the computation load. The other benefit is that it helps lift the performance (and accuracy) of models where there are many noisy features by eliminating such features. Feature selection generally looks to test several subsets of features and find the subset that minimizes the error rate.

Three types of feature selection methods exist:

1. ***Filtering:*** selects features as a pre-processing step. Uses a proxy measure such as inter/intra class distance instead of the error rate to score a feature subset.

2. ***Wrapper method:*** Uses predictive methods as a black box to score subsets of features. Each new subset of features is used to train a model and then test the model on a different subset of features. The error rate is the score of that subset.
3. ***Embedded method:*** Performs feature selection as part of the training process of the prediction method. It uses the LASSO method for constructing the model and shrinking regression coefficients to zero. It selects features that provide a non-zero regression coefficient.

There is also a procedure known as *feature extraction*. Feature extraction attempts to create new features from functions of the original features, whereas feature selection provides a subset of the features. Feature extraction has many applications. It's often used in data where there are many variables but small samples of data. Some of the use-cases for feature extraction include: analysis of written texts where we want to extract opinions and DNA microarray data, where we want to extract the pertinent DNA sequences.

6.44 Bridging Studies

Bridging studies are designed to "bridge" the gap between varied data from different studies. They have many applications in bridging gaps among randomized clinical trials such as the gap between efficacies and outcomes of drugs among different populations. Usually the population difference is ethnicity but it may also simply be geographical location. The studies are important for both pharmacodynamic and pharmacokinetic reasons.

Geographically separate populations have often evolved slight differences in receptor and enzyme makeup, this can lead to a drug having a drastically different binding affinity in one population to another.

6.45 Signal Boosting and Bagging Methods

We've already discussed signal boosting and bagging methods in previous sections. *B*ootstrap *agg*regating, as the acronym shows, is also known as the Bagging method. Bagging boosts the contribution of signals in machine learning and predictive models.

Bagging replaces a sample of K data elements with a sample from the set of observations and measures to increase (boost) results. It repeats the process to form a set of predictors. Bagging uses the ensemble of predictors by averaging the outputs of all observation sets. It's often used in regression trees and neural networks.

Boosting uses different samples of data as training sets. Then compares the lift (or reduction) in accuracy across multiple sample sets. Next it computes weights for samples to increase or decrease their membership in the training set. This step forms the ensemble which increases overall accuracy of the model.

6.46 Generalized Estimating Equation (GEE) Method

Generalized estimating equation (GEE) is a semi-parameterized regression technique to estimate average response over the population. It uses mean and variance of data for computation. The GEE method is commonly used in epidemiological studies and multi-site cohort studies since it can handle many types of unmeasured dependence between outcomes. It's the ideal alternative for compensating missing data in data set.

GEE method is not just an ideal method for repeated measures. It's appropriate for any situation where dependencies arise in the data (e.g. studies across families, clinics, etc.). It uses weighted combinations of observations to extract the appropriate amount of information from correlated data. GEE method relies on the independence across subjects to estimate consistently the variance of the regression coefficients.

6.47 Q-Q Plots

The quantile-quantile (Q-Q) plot is a graphical technique to determine if two data sets come from populations from the same distribution. It plots the quantiles of the first data set against the quantiles of the 2nd data set. It draws a 45-degree reference line. If the two quantiles plot along this line the two data sets come from the same distribution, otherwise they represent different populations. For example consider the Q-Q plot in the next figure.

In Figure 6.15, the two data sets don't come from populations with common distribution since the plot deviates from the 45-degree reference line.

6.48 Reduction in Variance (RIV) —Intergroup Variation

An RIV statistic is used frequently in healthcare to measure the explanatory power of casemix systems, i.e. the proportion of total length of stay (LoS) variation explained by the groups. A value of 0% means that the classification explains none of the variance in the dependent variable (e.g. LoS or cost), while 100% means it explains all of the variance. 100%, while theoretically possible, would suggest that all the data in each group have the same LoS/cost ratio. Typical results for LoS would be 30-40% while cost would be 60–70%.

The RIV, often expressed as R^2 to describe the predictive validity of the classifications, is calculated to describe the explanatory power of the grouping classifications.

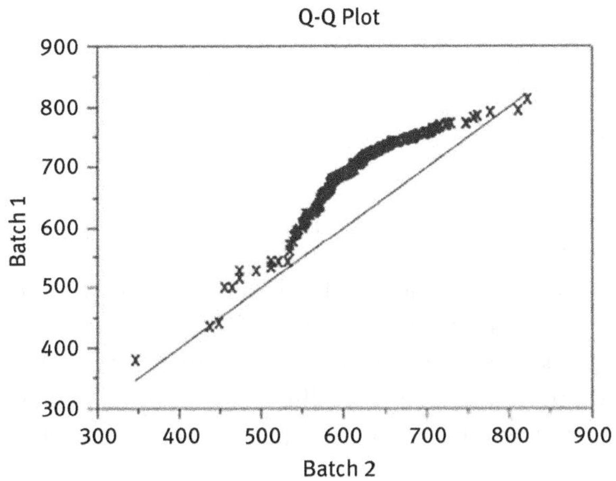

Figure 6.15: Example of Q-Q plot.

The unadjusted form of the calculation of RIV is the inverse of the ratio of the whole sum of squares (WSS) and the total sum of squares (TSS), expressed as a percentage.

$$R^2 = 1 - \frac{WSS}{TSS}$$

Where WSS = whole sum of squares, and
TSS = total sum of squares

$$WSS = \sum_{j=1}^{k} \sum_{i=1}^{n_j} (x_{ij} - \bar{x}_j)^2 \quad TSS = \sum_{j=1}^{k} \sum_{i=1}^{n_j} (x_{ij} - \bar{x}_j)^2$$

Where k = the number of groups
n_j = the number of cases in group j
x_{ij} = value of case i in group j
$\bar{x}_j$ = mean of group j
$\bar{x}$ = overall mean

6.49 Coefficient of Variation (CV)—Intra Group Variation

While the RIV statistic gives a result to be applied across groups, a statistic is required to measure the within-group variability or homogeneity. The ratio of standard deviation (SD) to the arithmetic mean of a group, or CV gives a measure of the relative variability within a single group.

$$CV = \frac{SD}{\bar{x}} = \frac{\sqrt{\sum_{i=1}^{n}(x_i - \bar{x})^2 / n}}{\bar{x}}$$

Where SD = standard deviation of the group
$\bar{x}$ = mean of group
x_i = value of case i in the group
n = the number of cases in the group

The CV is reported for a group to describe its homogeneity. A value of 0 would indicate that a group has no variance from the mean (i.e. standard deviation is equal to 0), while a CV value for a group above 1.00 would indicate heterogeneity within the group, where the standard deviation is greater than the mean. Caution should be used when the mean is close to zero as CV may be sensitive to small changes in the mean.

Chapter 7
Ensemble of Models: Data Analytics Prediction Framework

The data analytics model presented in this framework provides a feed-forward model to make predictions based on deep learning (machine learning) models trained on prior data. A prediction is a form of speculation about a state or outcome in the future. A *prediction* is foretelling an event or outcome when the ingredients for that event are in place. A predictor, also referred to as a *marker* is a variable that has predictive power and precursor correlation with an outcome.

7.1 Ensemble of Models

Since data changes from time to time or under different situations, using a single model is often inadequate to achieve a high level of accuracy and resiliency to changes in data. Hence, I've proposed using an ensemble of models and an overseer program (which I call an oracle) to provide a composite set of results from multiple models.

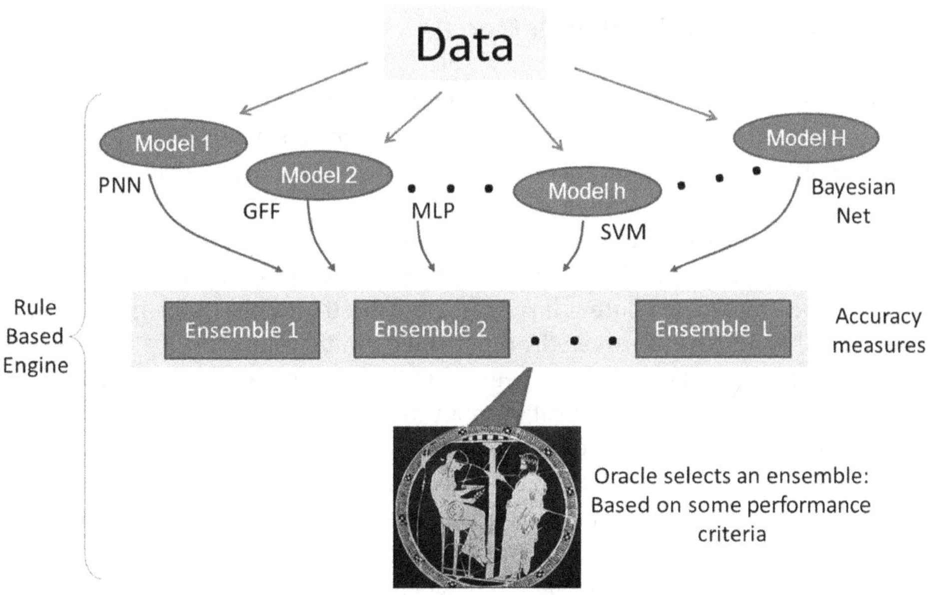

Figure 7.1: The ensemble (committee of model) framework for a robust data analytics system.

https://doi.org/10.1515/9781547401567-008

Given different data sets with different data sizes and data characteristic, each model can perform better (with more accuracy and specificity as will be explained later). This framework is the ensemble (also known as the committee of models) program. Figure 7.1 illustrates this framework.

In healthcare, the precursor to a disease is known as a *risk factor*. Thus, the spectrum of medical predictions starts with risk factors, leading to prediction, and then on to markers and finally to the occurrence of the disease or medical event itself. Figure 7.2 illustrates the chronology of events and progression of the individual's health status from risk factors toward a confirmed stage of disease or medical event manifestation.

The distance between time ticks are arbitrary and vary among individuals. The medical prediction models must take into account the prior history, risk factors, markers and the medical intervention as inputs to the model.

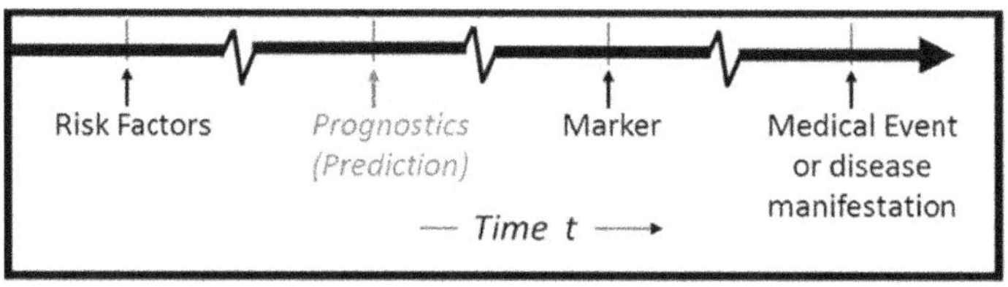

Figure 7.2: Progression of individual health condition.

7.2 Artificial Neural Network Models

The power of artificial neural networks comes in its ability to detect patterns, including those complicated situations when the traditional statistical analysis would take an inordinate amount of time that would render them impractical (Monterola, Lim et al. 2002).

Neural networks are important components to machine learning, deep learning and artificial intelligence. As a minimum, they consist of an input layer that receives dataset inputs, and output layer and produces the results of the network and at least one hidden layer between the input and output layers.

An artificial neural network is a network of interconnected processing elements that can classify patterns from a set of input data. Unlike the traditional computer architectures, known as von-Neumann computers, ANNs are trained, rather than programmed. When a set of data is fed into an ANN model, if there are patterns in the data, the ANN "learns" those patterns. Once the pattern is learned, the ANN model can classify a new set of data into the appropriate categories.

The suitable predictive mathematical model must offer accuracy and simplicity to learn from prior cases and easily be extensible to apply new data to make predictions. I use four different ANN algorithms to illustrate the concept of ensemble or committee of models. The four ANN algorithms selected in this ensemble are established through the literature among the most commonly used and accurate neural network models for prediction and classification. But, what improves the power and accuracy of prediction is not by the individual ANN models, but by an ensemble (committee) of them. Each model has certain relative strengths and weaknesses depending on the input data and computational constraints (Principe 2011). The ensemble framework improves each model's performance by collectively using them together.

To construct an ensemble, you can start with two models and keep adding as you develop new models. In general, you can utilize additional and different models ranging from random forest, to naïve Bayesian networks or principal component analysis and support vector machines. I used four ANN models for the case study which are listed below. The four models are:

1) PNN – Probabilistic neural networks are four layer networks. They classify data in a non-parametric method and are less sensitive to outliers in data. These models are known for performing well when datasets are small.
2) SVM – Support vector machine networks. SVM performs classification by constructing a two-layer network that defines a hyperplane that separates data into multiple classifications. This method is generally regarded among the more accurate classification models. Support vector machines are not regarded as neural networks, but they can be used as a solver method in a neural network model.
3) MLP trained with LM – Multi-layer perceptron with the Levenberg-Marquardt algorithm, is a gradient descent approach with variable step modification. This algorithm is regarded as a computationally efficient method.
4) GFN (Generalized feed-forward network) trained with LM – Generalized multi-layer feed-forward network with Levenberg-Marquardt algorithm. These models typically perform well when datasets are large and many data cases are available.

In order to make predictions on time-series data, a time-lag recurring network variation may be used instead of the above algorithms. The time-lag recurring network is essentially a time-series modeling approach that shifts the prediction several iterations forward in time and provides results of several samples ahead.

Different neural network models use different learning rules, but in general they all determine pattern statistics from a set of training examples that then classify new data according to the trained rules. Stated differently, a trained neural network model classifies (or maps) a set of input data to a specific disease from a set of diseases.

7.3 Analytic Model Comparison and Evaluation

Feasibility and utility of a model is gauged against five criteria of accuracy, well-posedness, utility, adaptability and economy. Each criterion is explored further in the following sections.

Accuracy

Accuracy of a model is the degree of closeness of the model's results to the system's actual value. Precision of a model is the degree to which repeated runs of the model under unchanged conditions produce the same results. Since the model is trained on retrospective data, it's easy to evaluate the prediction accuracy of the model to the actual clinical outcomes from prior retrospective patient cases.

Analysis about accuracy will include measurements such as calibration (agreement between predicted probability and observed outcome frequencies), discrimination (for example, the ability to distinguish between patients with and without the disease), sensitivity (proportion of patients who are correctly diagnosed as having the disease), specificity (proportion of healthy patients who are correctly diagnosed with negative results), likelihood ratio (LR, how much the odds of disease change based on a positive or negative test result) and receiver operating characteristic (ROC, a plot of sensitivity vs. one minus specificity) curves.

This framework uses an oracle program with several selection schemes to select the best combination of models in an ensemble that provide higher accuracy of prediction. a

Comparing prediction accuracy of ANN and other statistical models requires standards for comparison using classification performance indices. These indices include receiver operating characteristic (ROC, a plot of sensitivity vs. one minus specificity) curve, Area under Receiver-Operating Characteristics (AUROC, an overall measure of accuracy that measures the area under the ROC curve and where bigger area indicates higher accuracy), sensitivity, specificity, accuracy and positive predictive value (PPV, probability that someone with a positive test result to actually have the disease) and negative predictive value (NPV, probability that someone with a negative test result to actually not have the disease) (Bourdes et al. 2011).

Sensitivity measures the fraction of positive cases classified as positive. Specificity measures the fraction of negative cases classified as negative.

AUROC is a good overall measure of predictive accuracy of a model. It represents the area under the ROC curve, a measure of how well a model can distinguish between disease and normal groups. An AUROC value near 0.50 suggests no discrimination, namely one can flip a coin to decide. But, an AUROC close to 1.0 is considered an excellent discriminator (Linder et al. 2004). The single measure for accuracy comparison of models and committee of models will be Area Under Characteristic (AUC) curve.

Well-posedness: Model Stability and Immunity

In this section, I explain model stability. Stability refers to how immune a model is to small changes in data. Models can exhibit different well-posedness characteristics.

Generally a mathematical model is regarded well-posed if it meets Hadamard's three criteria:

1) The model has a solution
2) The solution in unique
3) The solution depends continuously on the data (Lucchetti 2006)

Conversely an ill posed mathematical model has initial, or boundary data, where an infinitesimal perturbation can grow unbounded away from the unperturbed solution. Generally, if it can be proven that the solution is uniformly bounded everywhere then it is well posed. But on the other hand it's possible to have unbounded solutions which are not ill posed. Since most classification problems include local optima as possible solutions, they're not regarded as well-posed.

However to overcome this limitation, you may consider genetic algorithms where the model exercises multiple optima and avoids getting trapped in a local optima. Therefore some level of regularization is necessary for the other models. In other words, one must know how to use additional assumptions to create a well-posed behavior in their models.

Utility: Practicality and Use-case

Once the model is trained, the model can be set up to run automatically at certain time intervals ranging from every minute to every several hours. It's practical to have all models trained in advance and run them in parallel. The oracle (overseer) program can provide the most accurate prediction by polling all models. The results of each model can be filtered and weighted by the oracle program to maximize true positives (TP) or minimize false negatives (FN) and other accuracy or precision targets.

Adaptability: Ability to Handle New Data Values and Data Types

ANNs are excellent machine learning tools, capable of adapting to new data sets, additional variables and all data types. However, the performance of machine learning models, such as accuracy, precision, specificity and sensitivity degrade over time. It's recommended that an ANN model be retrained after every few months over again in order to adapt to new data to overcome degradation and changes in temporal, environmental and demographic data. In the Ensemble framework, the historical data must be retained for future machine learning training and validation. ANN models can be re-trained as new data become available from the historical data sets.

Economy: Cost of Computation and Timeliness of Prediction

In computing, the computational cost of algorithms is determined by an asymptotic number of computations required for an algorithm to complete. The complexity measure of neural network algorithms provides an upper limit on the worst case scenario when the input variables and number of cases grow large. The computational cost of an algorithm is a function of number of steps to compute (time complexity), memory size (space complexity) and length of algorithm.

In ANNs, the number of computations is a time complexity, the number of *perceptrons* as a measure of space complexity and the number of weights as a measure of algorithm length. The complexity of ANNs has been shown to be NP-complete, namely given enough information and hardware; they can predict any input-output function in a finite time (Kon and Plaskota 2000).

The four classification and prediction models used in the case study were trained in under 5 minutes of CPU time and under one hour elapsed time on an average personal computer. The cost of computation is not extreme for a new prediction and can be completed within less than 2 minutes for cases of comparable dataset sizes, such as the case studied in this framework. This is to show that many of the algorithms can be trained quickly and many data analytics studies are within the realm of computational power available.

The development of the four models and their mathematical equations are covered in the next chapter.

Chapter 8
Machine Learning, Deep Learning—Artificial Neural Networks

8.1 Introduction to ANNs

Neural networks are critical tools for machine learning and artificial intelligence. They have a remarkable ability to capture the non-linear nuances of real world data. Finding patterns in data, data mining, classifying data, computer vision and speech recognition are among many problems that neural networks are able to solve.

Neural networks have been successfully applied to classify patterns based on learning from prior examples. Different neural network models use different learning rules, but in general they determine pattern statistics from a set of training examples and then classify new data according to the trained rules. For example, ANNs can be used to assist in clinical diagnosis. In such a use-case, a trained neural network model classifies (or maps) a set of input data to a specific disease selected from a set of possible diseases. Neural Networks represent an important class of algorithms in machine learning and in particular in deep learning.[57]

Artificial Neural Networks (ANNs) are inspired by biological learning processes. ANN models are parallel information processing constructs that attempt to mimic certain biological neural systems. ANN offers many advantages: it can model both linear and non-linear problems. It can scale up and down depending on the size. Their parallel construct provides self-healing and redundancy. Models based on ANN constructs attempt to answer several questions about learning, classification and pattern recognition. These attributes are useful features of cognition and reasoning that occur in many applications of decision making.

Artificial neural networks are used in a wide range of applications, from customer churn prediction and insurance claim prediction in the insurance industry to character recognition, face recognition and cancer detection in other use-cases. Other applications include predictive modeling, recommender systems and data mining where finding patterns in data are needed.

The goal of ANN is to mimic the nervous system. Just as the nervous system consists of an interconnection of simple units, called nerve cells, an ANN consists of many independent but inter-related elements (called neurons) organized into layers. Each neuron transmits an excitation or inhibitory signal to another neuron. The contribution of the signals depends on the strength of the synaptic connection. Similarly, biological neural learning happens by the modification of the synaptic

57 As we'll study later, a deep learning model is essentially a neural network with many hidden layers.

https://doi.org/10.1515/9781547401567-009

strength. In a neural network the synaptic strengths are represented by weights associated with each input.

A typical ANN is composed of layers connected to each other by full or random connections. There are typically two layers with connections to the external world: an input layer where data is collected and an output layer that presents the outcome or response of the network. But multi-layer ANN models are common. Figure 8.1 shows a simple neuron consisting of input signals designated by $x_1, \ldots, x_k$, weights associated with each signal $w_{i1}, \ldots, w_{ik}$, a summing junction and an activation function that produces output Y_i.

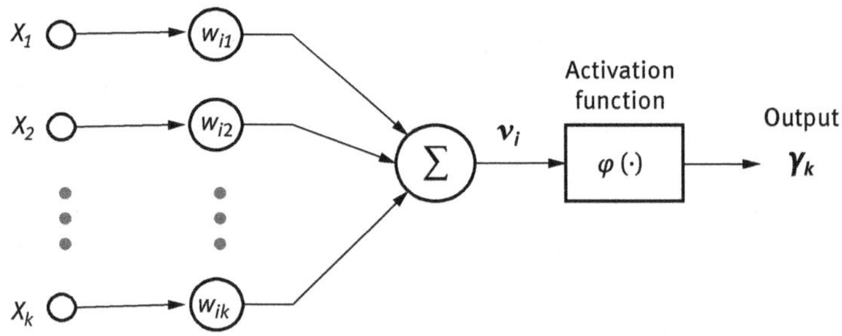

Figure 8.1: A simple neuron with activation function for node i.

For each neuron, the summation function aggregates a weighted sum of inputs while the activation function transforms the sum into the final output of the neuron. The formula of each step is shown below:

$$v_i = \Sigma_{j=1}^{k} x_j w_{ij} \tag{10}$$

$$Y_i = \varphi(\star) \tag{11}$$

In Figure 8.2, the general structure of a multi-layer ANN is shown. The data gathered about a patient's condition is fed into the model through a layer of neurons. Here four input signals are shown. The result of each layer is an activation function whose output is input to the next layer. There are m rules in the framework, each detecting a particular disease. Layers are shown by neurons q_{jpm} and weights that connect layer $(p+1)$ to layer p by w_{ijpm}, where i denotes the number of neurons in layer p and j represents the number of neurons in layer $(p+1)$; the subscript m denotes the perceptron parameters for model m.

Every line connecting two neurons together has a weight associated with it. (In Figure 8.2 only the weights of outer connections are shown to keep the diagram readable.) The equation to calculate the value of every neuron in each layer in the n-layer network above can be described as:

$$q_{i(p+1)m} = \Sigma_{j=1}^{k-p+1} q_{jpm} w_{ijpm}, \text{ for } p = 1, \ldots, n-1, \text{ and layers, } i = 1, \ldots, k-p \tag{12}$$

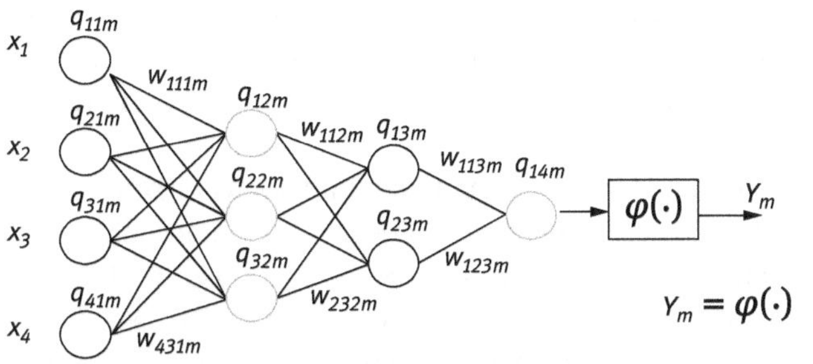

where q_{ipm}, for $p = 1$ to n, $n = 4$ is the neuron no. i at layer p for model m, and w_{ijpm}, for $p = 1$ to n-1, $n = 4$ is the weight of neuron j to neuron i in layer p for model m

Figure 8.2: A 4-layer neural network.

where k is the number of input measurements $x(t)$. Some well-known documented advantages of ANN are learning and pattern recognition. Depending on the activation function, the final output can be a "1" or "0" indicating whether the patient is in danger of developing DVT/PE symptoms or not. Among the training methods, back-propagation is a common technique for training neural networks. This framework used this technique to train a model and applied it to a new set of patient data for predictive purposes.

Back-propagation consists of two steps: In step one; the researcher calculates error contributions to the response function Y. This step computes how much each neuron has contributed to the total error in the response value. Error is defined as the difference between the ideal (or expected) result versus the actual response value. Neurons with higher weights have contributed more to the total error and therefore their weight needs to be adjusted more. In step two, the algorithm adjusts the weights starting from the outer layer neurons, going back to the hidden layers, and finally reaching the weights of the input layer. When this algorithm completes, the network has been trained. This is called supervised learning because it defines the ideal (or expected) response value.

Once the neural network model is trained, it can be applied to a fresh or incomplete set of data. The outputs will provide predictions based on the inputs and adjusted weights.

8.2 A Simple Example

The following are two examples of simple, single-layer perceptron classification. The original examples appear in Zurada (Zurada 1997), Haykin (1998), Sengupta (2009), and Masters (1995). We're going to predict which patients will develop a disease known as DVT by analyzing a broad range of health data for each individual.

Input data about patients can be classified into two categories of predictions: DVT-True and DVT-False, by looking at prior patient data. The objective of the single-layer perceptron is to determine a linear boundary that classifies the patients on either side of the linear boundary.

As shown in Figure 8.3, the intent is to classify patients into two categories separating by a boundary called a *decision boundary line*. A linear set of equations define this boundary. The region where the linear equation is >0 is one class (DVT-True), and the region where the linear equation is <0 is the other class (DVT-False). If the linear equation is = 0, then the patient falls on the line. Our goal is to devise the line such that patients are divided in either side of the line. The line is defined as:

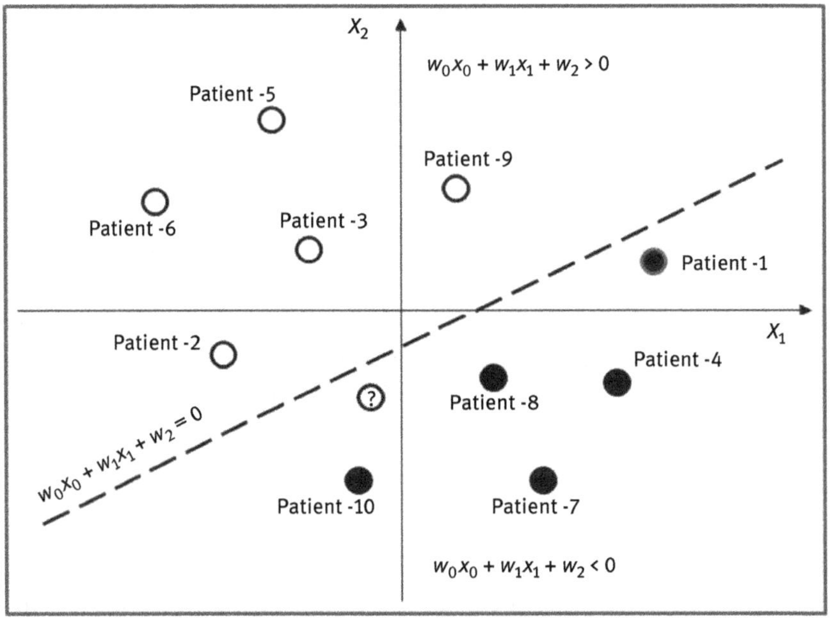

Figure 8.3: Classification using single-layer perceptron.

$$w_1x_1 + w_2x_2 + w_0 = 0$$

You can apply a threshold function to classify patients based on the following threshold function:

$$p(x_1, x_2) = \begin{cases} 1 & \text{if } w_1x_1 + w_2x_2 + w_0 \geq 0 \\ -1 & \text{if } w_1x_1 + w_2x_2 + w_0 < 0 \end{cases}$$

Suppose we're considering classifying patients by only four input variables, glucose (G), body mass (M), systolic blood pressure (S) and white blood cell count (B), represented by x_1, x_2, x_3, and x_4. The threshold function would be computed as follows:

$$p(x_1, x_2, x_3, x_4) = \begin{cases} 1 & \text{if } w_0 + \sum_{i=1}^{4} w_i x_i \geq 0 \\ -1 & \text{if } w_0 + \sum_{i=1}^{4} w_i x_i \leq 0 \end{cases}$$

Let's assume the following weights and input values for the classification example are given as shown in Table 8.1. We can assume the disease under study is DVT.

Table 8.1: Computing classification using single layer perceptron classification.

Weights	Values	Inputs	Values
w_1	2	x_1	-1
w_2	0	x_2	2
w_3	3	x_3	0
w_4	-1	x_4	-4
w_0	1	*Bias*	1
$p(x_1, x_2, x_3, x_4) =$ $2^* - 1 + 0^*2 + 3^*0 + -1^* - 4 + 1^*1 = 3$ $\Rightarrow$ class $= 1$ or DVT – True			

If this classification is incorrect, then it's necessary to adjust the weights and repeat the process until the patient is correctly classified. Suppose the correct classification is (–1), then the calculation proceeds as shown in Table 8.2. The results indicate

Table 8.2: Revising weights to correct misclassification.

Weights	Values	Inputs	Values	New Weight calculation when actual *class* = –1
w_1	2	x_1	-1	$w_1 = w_1 + class * x_1 = 2 + (-1) \star (-1) = 3$
w_2	0	x_2	2	$w_2 = w_2 + class * x_2 = 0 + (-1) \star 2 = -2$
w_3	3	x_3	0	$w_3 = w_3 + class * x_3 = 3 + (-1) \star 0 = 3$
w_4	-1	x_4	-4	$w_4 = w_4 + class * x_4 = -1 + (-1) \star (-4) = 3$
w_0	1	*Bias*	1	$w_0 = w_0 + class * x_0 = 1 + (-1) \star 1 = 0$
$p(x_1, x_2, x_3, x_4) = 3^* - 1 + -2^*2 + 3^*0 + -1^*3 + 0^*1 = -7 \Rightarrow$ class $= -1$ or DVT – False				

which class the data belongs to, which in this example the classification is no-disease or DVT-False.

8.3 A Simplified Mathematical Example

In this section, a simple ANN model is presented as an example using the XOR logic table for illustration. As shown in Table 8.3, the XOR table returns a value of 0 if both inputs are identical (both 0's or 1's) and returns a value of 1 if one or the other input is a 1.

Table 8.3: The XOR Logic Table and Results from ANN Model.

Input x_1	Input x_2	Ideal Value
0	0	0
0	1	1
1	0	1
1	1	0

It's possible to develop a 3-layer neural network to compute the result for each pair of inputs x_1 and x_2. A third input called *bias* is also introduced to construct the model. The value of bias is always 1. Bias is added to add flexibility to the behavior of the neural network. It's added to all of the input and hidden layers of the model. Adding bias is analogous to adding an intercept in a regression model. If a neural network does include a bias, for certain input values, it may not be able to produce a value other than "0" as its output.

In the first iteration, random weights are used. There are a total of nine weights in this model as shown in Figure 8.4. Simply put, the output is a function of weights and inputs. This is an example of supervised learning as the weights in the ANN model get trained to produce the desired output.

The goal is to adjust the weights iteratively until the ANN model produces the *ideal value*. In the first iteration, the model produces some results shown in the *output* column. The model computes the error as Mean Square Error (MSE) and uses the error to adjust the weights.

The iterations continue and weights get adjusted until the error term is below a threshold (in this case less than 0.009). Eventually the ANN model stops and the output is the *final* result as shown in the last column in Table 8.4. The computed results are close to the ideal values (close to 0 or 1), only different by a small margin of error.

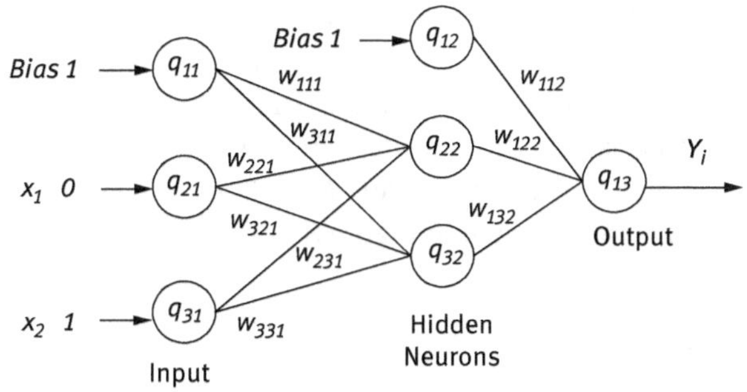

Figure 8.4: A 3-layer ANN network to compute the XOR logic table.

Table 8.4: The XOR logic table and results from ANN model.

Input x_1	Input x_2	Ideal Value	Output	$(Error)^2$	Final Results
0	0	0	0.2	0.04	.00875
0	1	1	0.3	0.49	.99130
1	0	1	0.4	0.36	.99123
1	1	0	0.5	0.25	.00568

8.4 Activation Functions

One of the key features of artificial neural networks is that a linear neuron output can map into a non-linear activation function (Haykin 1998, Sengupta 2009, Zurada 1997). Activation functions play an important role to transform the output of the perceptron and shape the result before it's fed into the next layer. The activation function may clip the value or transform it into a function. There are three types of activation functions.

Given inputs $x_0, x_1, \ldots, x_n$, the output v_k is the result of summation from Equation (10) and $Y_k = \varphi(\star)$ is the result of the activation function. The activation function results in an S-shaped curve known as the *sigmoid function.*

In the first type, as v_k changes from $-\infty$ to $+\infty$ the output can vary from 0 to 1, namely $y_k = [0, 1]$. This is the logistics function shown in Figure 8.5. The activation function for this type of neural network is shown as:

$$\varphi(\star) = \frac{1}{1 + \exp(-av)} \tag{13}$$

$$\varphi(\star) = \begin{cases} 1 & when\ v \rightarrow +\infty \\ 0 & when\ v \rightarrow -\infty \end{cases} \tag{14}$$

This is the simplest neuron formulation. It's possible to change the shape of the S-curve by changing the values of a. When a is small, the curve appears as a smooth function, meaning it is curved like an "S" as shown in Figure 8.5. But, when a is very large, this function approaches the threshold function, a model proposed by Pitts and McCollough (It's also known as the Pitts-McCollough model). In that case, the curve looks more like a step as shown in Figure 8.7.

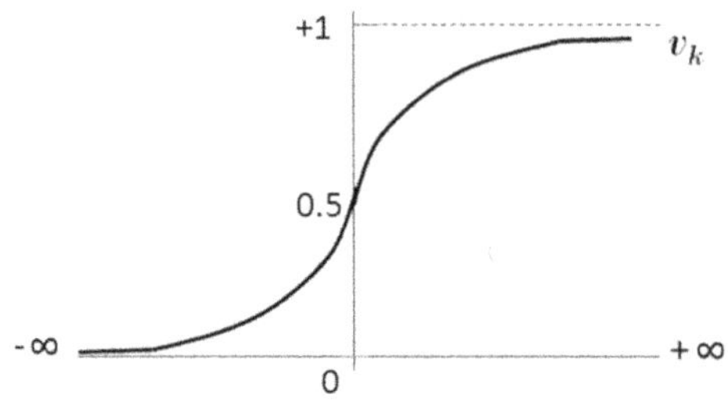

Figure 8.5: Sigmoid function for activation function v_k between [0,1].

The second type of activation function has a range from –1 to +1, shown with the following equation:

$$\varphi(v) = \tanh(av) \tag{15}$$

The tanh(.) is the hyperbolic tangent function computed as follows:

$$\tanh(v) = \frac{e^v - e^{-v}}{e^v + e^{-v}} \tag{16}$$

The activation function $\varphi(v)$ is determined according to the value of v:

$$\varphi(\star) = \begin{cases} 1 & when\ v \to +\infty \\ -1 & when\ v \to -\infty \end{cases} \tag{17}$$

The shape of S-curve representing this activation function is shown in Figure 8.6.

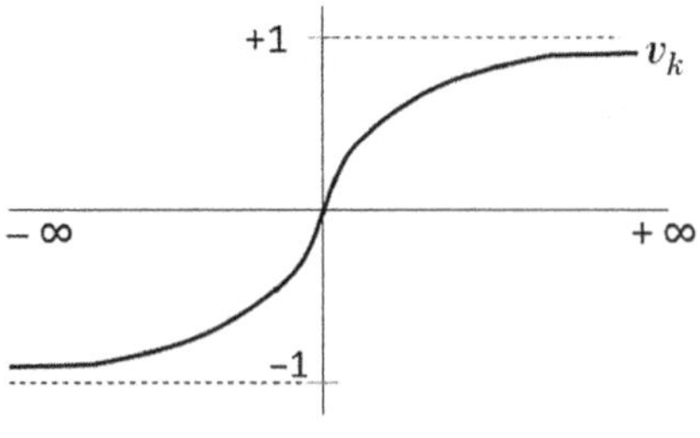

Figure 8.6: Activation function for v_k between [–1, 1].

The third type of activation function is the stochastic model determined by:

$$\varphi(v)=\begin{cases}1 & with\,prob\,p(v)\\ 0 & with\,prob\,1-p(v)\end{cases} \tag{18}$$

and

$$p(v)=\frac{1}{1+e^{(-v/T)}} \tag{19}$$

When $p(v)=1$ then $T=0$ and this activation function becomes a deterministic model. As T gets larger, there is more stochastic behavior in the model. To better illustrate the role of T, it's possible to think of it as temperature or kinetic energy, borrowing this concept loosely from the third law of thermodynamics. Figure 8.7 shows the S-curve associated with this activation function. The various S-curves illustrate the effect of T on the values on the curve.

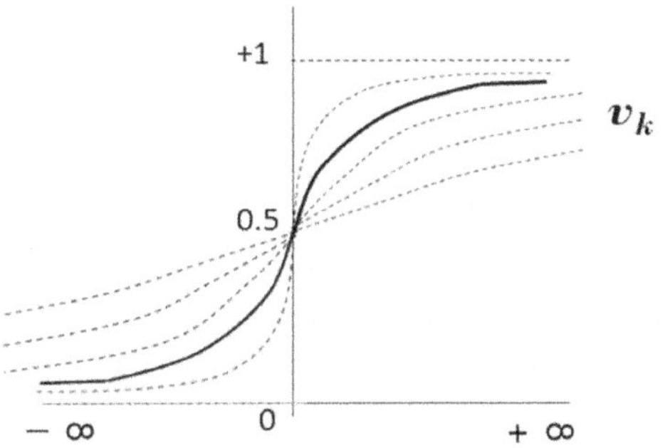

Figure 8.7: Stochastic activation function for v_k.

8.5 Why Artificial Neural Network Algorithms

Of the various statistical and computational methods covered in predictive models, Artificial Neural Networks offer unique advantages that make them a suitable tool for medical research and prognostics. These advantages outweigh some of the criticisms that have been leveled against ANNs (TU 1996).

The primary criticisms of ANNs include: a "black-box" approach to data, proneness to over-fitting and greater computational burden. Requiring greater computational power is less of an issue now as the desk top and portable computers have much more powerful computational power. The criticism about "black-box" approach is not a serious limitation in this framework since multiple models are used and supervised learning is applied.

Over-fitting is a weakness that occurs when a model is trained to a specific data set and performs poorly on other datasets not used in training. This weakness can be avoided through multiple iterations of cross-validation and setting aside a separate test data batch as employed in this framework. Additionally using multiple models can overcome the issue of one model getting over-trained by one data set versus another data set.

In contrast, the advantages and reasons for choosing ANNs as predictive models are significant considerations:

- Ability to model complex non-linear relationships between input data and output
- Ability to learn and adapt to patterns in data
- Resilience toward missing data elements
- Many algorithms are available to choose from
- Ability to handle a large amount of variables
- Ability to handle diverse types of data
- Ability to detect all possible interactions among predictor variables

8.6 Deep Learning

Artificial neural networks as we've seen consist of at least 3 layers: An input layer, one hidden layer and one output layer. The notion of deep learning refers to an artificial neural network model that has multiple hidden layers. Studies have shown that a multi-layer artificial neural network is capable of deep learning, namely is able to model any non-linear relationship in a system.

Adding more hidden layers can improve accuracy of the prediction, but only up to a point. Studies and research has shown that adding too many hidden layers (say beyond 5 hidden layers) causes the model to lose accuracy and become so "generic" that it can't distinctly distinguish between different input signals. In other words, as you build your model, you should experiment with one, then two and add more hidden layers until you find the optimum accuracy level.

8.7 Mathematical Foundations of Artificial Neural Networks

A mathematical foundation of ANNs is presented in this section using the common conventions of ANN formulations, so inputs are represented by x, the desired output by d, weights by w, and ANN's output by y (Zurada 1997, Haykin 1998, Wang 2012, Sengupta 2009). As an introduction, a simple neuron is presented followed by formulations for classification, memory and learning. A single neuron can be constructed with a single activation function. Consider finding a regression line for the histogram shown in Figure 8.8. The regression line is represented by

$$y = mx + b \tag{20}$$

Where m is the slope of the line and b is a constant, known as bias. The corresponding neural network representation uses a single neuron where the weight parameter w_{11} corresponds to slope m and w_{10} is equal to a constant 1.0.

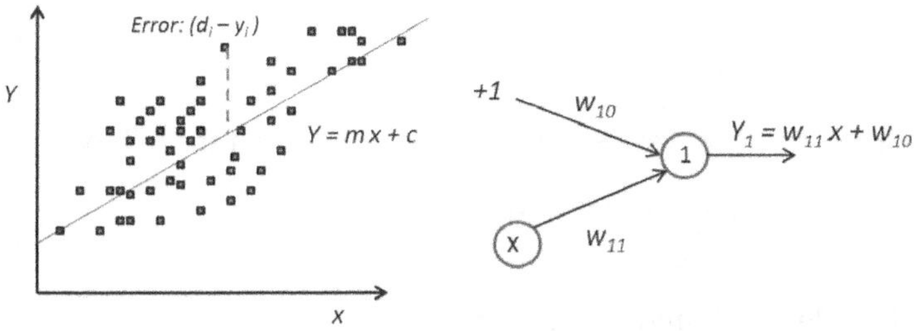

Figure 8.8: A regression line and its equivalent single neuron representation.

If y is dependent on multiple inputs x_j, one can think of bias as another input to the neuron with a weight w_{kO}. To find the best fitting line, the goal is to minimize the errors E, by adjusting weights. Given multiple x_j one can use the gradient descent method discussed below to find the minimum E and the corresponding weights.

8.8 Gradient Descent Methods

The *gradient descent method* is used as an iterative process to determine the weights associated with each input x. This method which is also known as *error correction* learning in ANN literature, works to minimize the total error as the method of training the model and determining the appropriate weights. The following derivation is adapted from Zurada (Zurada 1997) and Sengupta (Sengupta 2009), improved and revised specific to this framework. Total error E can be written as:

$$Total\ Error\ E = \Sigma_j\ E_j = \frac{1}{2}\Sigma_j(d_j - y_j)^2 \tag{21}$$

This represents the total error E for point j. The expression d_j is the target output (desired output) at point j, and y_j is the actual output at point j.

Let's now consider all possible outputs $y_O, \ldots, y_m$ where $0 \leq j \leq m$, and

$$\begin{aligned} y_0 &= f_0(x_1, x_2, \ldots, x_n) \\ y_1 &= f_1(x_1, x_2, \ldots, x_n) \\ &\ldots \\ y_m &= f_m(x_1, x_2, \ldots, x_n) \end{aligned} \tag{22}$$

Total error E is the combined error of all errors for outputs y_k. It's common to use ½ of the sum in (14) since as will be explained later, it makes mathematical manipulations easier as one takes the derivative of this term and the gradient will be multiplied by 2.

Let's define the gradient, namely the rate of increase for (i,j) pair connection as:

$$G = \frac{\partial E}{\partial w_{ij}} = \frac{\partial}{\partial w_{ij}} \Sigma E_j = \Sigma_j \frac{\partial E_j}{\partial w_{ij}} \tag{23}$$

Next, it's possible to apply partial derivatives and the chain rule to get the following:

$$\frac{\partial E}{\partial w_{ij}} = \frac{\partial E}{\partial y_j} \cdot \frac{\partial y_j}{\partial w_{ij}} \tag{24}$$

For sake of simplicity let's denote d_j and y_j as follows:

$$d_j = \Sigma d_j, \ \ and \ y_j = \Sigma y_j$$

You can take derivative of the total error, Equation (21) to get the following:

$$\frac{\partial E}{\partial y_j} = -(d_j - y_j) \tag{25}$$

$$y_j = \Sigma_j \, w_{ij} x_j \tag{26}$$

$$\frac{\partial y_o}{\partial w_{oi}} = \frac{\partial}{\partial w_{oi}} \Sigma_j w_{oj} x_j = x_i \tag{27}$$

$$\frac{\partial E}{\partial w_{ij}} = -(d_j - y_j) x_i \tag{28}$$

Where j is the output unit, and i is the input unit. Thus the derivative of error with respect to w_{ij} has been formulated. In order to move the opposite direction to the derivative, you can apply the corrections to the w_{ij}'s by multiplying a (−) sign to the difference. The (−) sign is applied because the goal is to minimize error. The correction can be written as:

$$\Delta w_{ij} = (d_j - y_j) x_i \tag{29}$$

The new synaptic weight will be computed using the following for several iterations until Δw_{ij} is less than a given threshold set by the user:

$$w_{ij(new)} = w_{ij(old)} + \Delta w_{ij} \tag{30}$$

It's possible to use η to represent the rate of descent in (29). So η is the learning rate that reduces total error E, with every iteration and can be defined by the researcher to regulate Δw_{ij}, the rate of descent.

8.9 Neural Network Learning Processes

There are five major categories of learning models in neural networks. One of these learning methods called error correction based learning was already discussed in section 8.8 under the gradient descent approach. The other four categories are:
- Memory based learning
- Hebbian based learning
- competitive learning and
- Boltzman learning model

These learning methods are adapted from Sengupta (2009), Zeruda (1997), Wang (2012), Masters (1995) and Haykin (1998). They're refined and revised for this framework and are included for the sake of completeness.

Memory Based Learning

Memory based learning works to retain the relationship between the input vector and output. Given input vector $\bar{x}$ defined by $\{x\}_{i=1}^{N}$, and desired output d_i, this association can be shown by the expression $\{x_i, d_i\}_{i=1}^{N}$. When the model is applied to a new pattern $\bar{x}_i$, since this pattern is initially unknown let's start with a test pattern $\bar{x}_{test}$ and find the Euclidean distance between $\bar{x}_{test}$ vector and the new pattern $\bar{x}_i$ vector. Let's assume that $\bar{x}'_N \in \{x_1, x_2, x_3, \ldots, x_N\}$ is the set of nearest neighbor points of $\bar{x}_{test}$ vector, then it implies that the distance of pattern $\bar{x}_i$ from $\bar{x}_{test}$ is minimum over the set of all $\bar{x}_i$. This can be shown by the following expression to be true for all distances over i:

$$\min_{(over\, i)} d\{x_i, \bar{x}_{test}\} = d(\bar{x}'_N, \bar{x}_{test}) \tag{31}$$

To improve this algorithm it's prudent to look at the nearest neighbors and find the set that offers minimum distances. Memory based learning is ideal for pattern recognition and classification of data as shown by the example in Figure 8.9. This approach helps

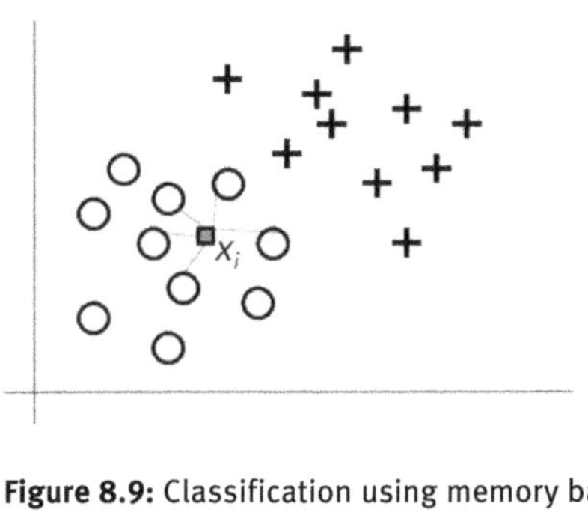

Figure 8.9: Classification using memory based learning.

keep outliers out of classification. This is the k-nearest neighbor classification. Let's suppose we want to classify a new data point. In Figure 8.9, the new point marked by x_i must be classified between the "+" or the "O" shapes. Since its nearest neighbors are the "O" shapes, it will get classified as a member of the "O" set.

Hebbian Based Learning

The goal of Hebbian based learning is to retain the association between the input vector and the output. This method is attributed to Donald Hebb, a neurobiologist who in 1949 introduced his theories of neuron adaptation in the brain during the learning process. In the Hebbian learning process, the amount of adjustment to weight w_{kj} is defined by:

$$\Delta w_{kj}(n) = f(y_k(n), x_j(n)) \tag{32}$$

The delta is the amount of change in w_{kj}. This is the adjustment in w_{kj} at time step n. The adjustment is expressed as a function of responses y_k and input x_j at time step n. You can re-write this expression in terms of pre-and post-synaptic responses:

$$\Delta w_{kj} = \eta y_k(n) x_j(n), \tag{33}$$

where η is the rate of learning. This expression is known as the *activity product rule*. It's important to note that η is constant. Assuming that one keeps x_j constant then it's possible to plot Δw_{kj} and $y_k(n)$ to get a line that intercepts through the origin with slope of $\eta x_j(n)$ as shown Figure 8.10. As y_k increases, so does Δw_{kj}. Eventually the synaptic weight reaches its saturation point where not more learning is possible.

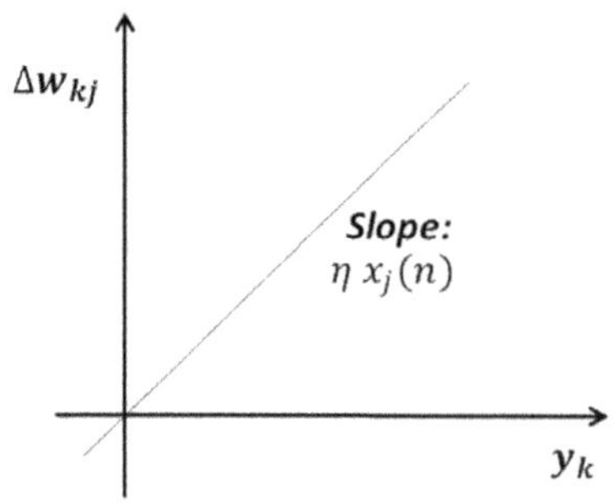

Figure 8.10: Slope of the activity product rule represents the rate of learning.

Let's define $\bar{x}$ and $\bar{y}$ as time averaged values of x_j and y_k. Then it's possible to define:

$$x_j(n) = x_j - \bar{x}, \quad \text{and} \quad y_k(n) = y_k - \bar{y}.$$

By definition of covariance, it's possible to write the change in weights w_{kj} as the covariance of distance of x_j and y_k from their respective time averaged values $\bar{x}$ and $\bar{y}$.

$$\Delta w_{kj} = \eta(x_j - \bar{x})(y_k - \bar{y}) \tag{34}$$

Since the average effect of change over the entire input values of x_j is desired, $\bar{x}$ can be recognized as a constant over the course of x_j. This relationship can be shown by a line that intersects Δw_{kj} and y_k as shown in Figure 8.11. This figure shows the relationship between Δw_{kj} and y_k for a given point x_j such that $(x_1 - \bar{x})$ is a constant.

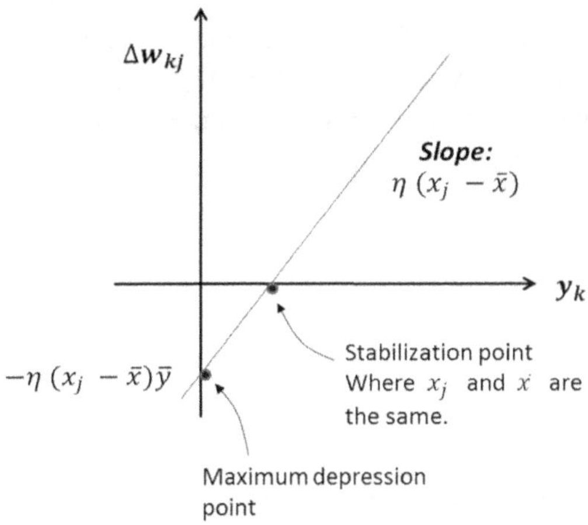

Figure 8.11: The covariance relationship between response and input.

When applying the covariance approach, three conditions are possible:

$$\begin{aligned}
&(i)\ \ w_{kj}\ \textit{increases if}\ x_j > \bar{x}\ \textit{and}\ y_k > \bar{y}\\
&(ii)\ \ w_{kj}\ \textit{decreases if either} \begin{cases} a)\ x_j < \bar{x}\ \textit{and}\ y_k > \bar{y} \\ b)\ x_j > \bar{x}\ \textit{and}\ y_k < \bar{y} \end{cases}\\
&(iii)\ w_{kj}\ \textit{increases if}\ x_j < \bar{x}\ \textit{and}\ y_k < \bar{y}
\end{aligned} \tag{35}$$

Competitive Learning

In *competitive learning*, each neuron competes to increase its response value while minimizing the other neuron's output. The winning neuron will be preferred in future iterations of learning. The mathematical model of competitive learning is based on:

$$y_k \begin{cases} 1\, if\ v_k > v_j,\ for\ all\ j\ when\ j \neq k. \\ 0\ otherwise \end{cases} \tag{36}$$

The sum total of all weights are set to 1 for all k:

$$\Sigma_j w_{kj} = 1,$$

For example, consider three clusters of input variables as shown in Figure 8.12. One can write x_j as a vector $\bar{x} = [x_1, x_2, x_3]$. If the relationship $\|\bar{x}\| = 1$ is enforced, namely that if it's required that $\sqrt{x_1{}^2 + x_2{}^2 + x_3{}^2} = 1$, then there can be several vectors $\bar{x}_1$, $\bar{x}_2$, $\bar{x}_3$ to represent different patterns as shown in Figure 8.12. The goal is to classify data into any one of n patterns. The clusters of patterns are grouped into a set of data in vectors (in this example in vectors $\bar{x}_1$, $\bar{x}_2$, $\bar{x}_3$) such that: $\Sigma_j w_{kj} = 1$, for all k. In addition, it's possible to show weights for each cluster as a vector. For example, the vector of weights for the first cluster can be shown as: $\bar{w}_1 = [w_{11}\ w_{12}\ w_{13}]$.

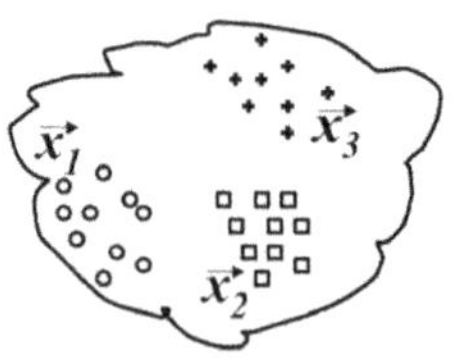

Figure 8.12: Using competitive learning to classify data into different patterns.

In this competitive learning model, typically the most central element (or neuron) in a cluster is the winner and all other weights conform (or align) to it. The competitive learning rule is that Δw_{kj} is determined by:

$$\Delta w_{kj} = \begin{cases} \eta(x_j - w_{kj}) & if\ neuron\ k\ wins\ the\ competition \\ 0 & if\ neuron\ k\ loses \end{cases}$$

Boltzman Learning Model

The Boltzman learning process is derived from statistical mechanics and mimics a stochastic learning model. In this model, the neurons constitute a recurrent structure that allows self-feedback. The neurons take a binary value of +1 or −1. The model is represented by:

$$E = -\frac{1}{2} \Sigma_j \Sigma_k w_{kj} x_k x_j \quad \text{where} \quad j \neq k$$

The visible neurons are the output layer. The inner neurons as shown in Figure 8.13 are hidden neurons. This model is regarded as stochastic as a change in one outer neuron changes the value of E. The probability that a neuron x_k flips it state from one state to another state is defined by:

$$P(x_k \rightarrow -x_k) = \frac{1}{1 + \exp(\frac{-\Delta E_k}{T})}$$

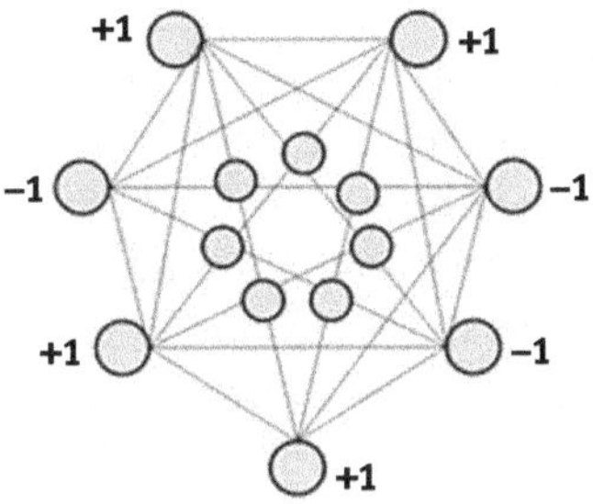

Figure 8.13: Inner and outer neurons in a Boltzman learning model.

The change in E_k from a flip is denoted by ΔE_k. If E is be regarded as an energy function, then ΔE_k is the change of energy from a flip of state in a neuron. The variable T is the pseudo temperature representing the level of noise or stochasticity. In a Boltzmann network, a neural can be either in a clamped condition or a free condition. The clamped condition for a neuron means that its value is set and not changeable. The free condition means that the value of the neuron can change during that state of operation. Let's assign two variables:

P_{kj}^{+}: the correlation between neuron k and neuron j in the clamped condition
P_{kj}^{-}: the correlation between neuron k and neuron j in the free condition
Then the Boltzman learning rule is defined by:

$$\Delta w_{kj} = \eta(P_{kj}^{+} - P_{kj}^{-}) \quad \text{where} \quad j \neq k.$$

8.10 Selected Analytics Models

Since the ensemble framework uses multiple models for classification and prediction, it's important that close attention is given to accuracy of each model. I selected the following four models for the ensemble because each provides certain strengths and characteristics that make it appropriate for certain types of data and prediction. It has also been established that these models are among the most accurate neural network models for classification. Below is a summary of advantages and disadvantages of each ANN method (Masters 1995):

1) PNN – Probabilistic neural networks are four layer networks. They classify data in a non-parametric method and are less sensitive to outlier data. It's been demonstrated that probabilistic neural networks using only four layers of input, pattern, summation and output perceptron can provide accurate and relatively faster classifications than the back-propagation neural networks (Principe, Euliano, Lefebvre 1999).
2) SVM – Support vector machine networks. SVM performs classification by constructing a two-layer network that defines a hyperplane that separates data into multiple classifications. The SVM is a non-probabilistic binary linear classifier. It takes a set of input data and determines which of possible classes the input is a member of. SVMs are ideal for classification and clustering data elements.
3) GFN (Generalized Feed-forward) trained with LM – A feed-forward neural network consists of one or more layers of nodes where the information flows in only one direction, forward from the input nodes and there are no cycles or loops in the network. In the multi-layer model, each node has a direct connection to the nodes in the subsequent layer. The sum of products of the weights and the inputs are calculated in each node (Haykin 1998). I use the Levenberg-Marquardt (LM) algorithm as the core method in this GFN model. I'll explain the Levenberg-Marquardt algorithm in the following sections.
4) MLP trained with LM – Multi-layer perceptron, a method similar to gradient descent approach with variable step modification. Several variations of this model have been proposed, including the Levenberg-Marquardt model (Wilamowski & Chen 1999) which is known to be among the most efficient algorithms.

The next section is a more in-depth mathematical review of the four neural network approaches used in this book.

8.11 Probabilistic Neural Networks

Recall the classification problem from Figure 8.3 where the goal is to classify an unknown patient (shown by the symbol "?") into one of the two groups. The most straightforward method would be to check the distance from the nearest neighbor. But this method, while simple, has weaknesses. It can be misclassified into one group when in fact it belongs to another groups' cluster. The goal is to define a "sphere of influence" function to represent the spread of distance separating an unknown point from a training set point.

Think of this "sphere of influence" as a function that visually resembles an "umbrella" as shown in Figure 8.14. Such a function would have a peak at zero distance from the training set point and taper off to zero as the distance away from the training set point increased. A proposed classifier would compute the sum of this function for all training set points of each population and classify the unknown into the population that has the greatest sum.

The following derivation is adapted from Sengupta (Sengupta 2009) and Zurada (Zurada 1997), improved and revised specific to predictive modeling. A mathematical construct that can help define such a function is the Gaussian function:

$$f(x) = ae^{-\frac{(x-x_i)^2}{2\sigma^2}} \tag{37}$$

Where a, x_i, σ are > 0, and a is the height of the curve's peak (or amplitude), x_i is the position of the center of the peak and σ is the width (or the spread) of the curve. Hardy (2006) has illustrated a 2-dimensional graph for x_0, x_1 as shown in Figure 8.14, where the values of x_0, x_1 are set to origin (0,0). Coefficients a, and σ can take any positive values. The scaling parameter σ controls the width of the area of influence.

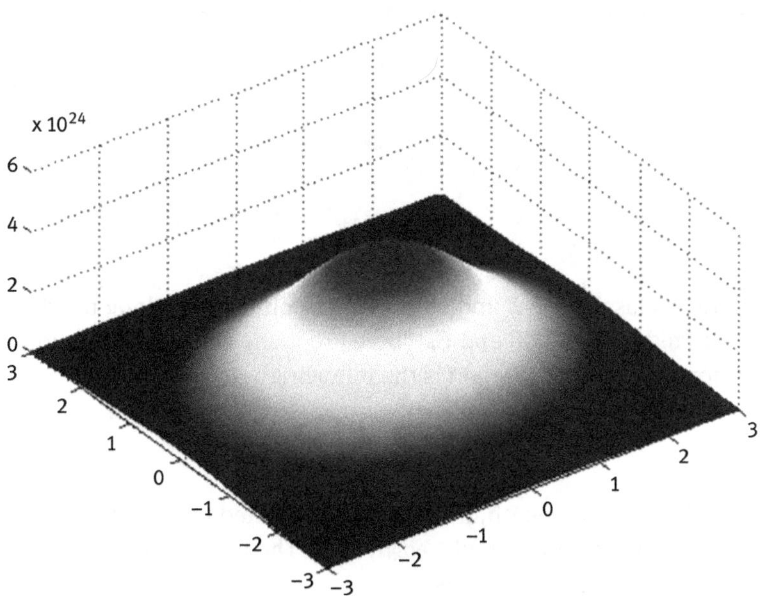

Figure 8.14: Graph of a 2-dimensional Gaussian function.

The idea behind probabilistic neural network (PNN) is that each training element represented by a Gaussian pattern unit, adds to the likelihood that nearby data has the same classification. To compute the classification of a data point, let's calculate the response for the point within every category and select the category that has the highest response. Each trained data point corresponds to a pattern unit that is a Gaussian function with its peak centered on the parameter's location.

The idea behind this classification approach is that for a new data point, we measure the average "distance" between the new point and all other points in the classification. The smaller the average "distance" of a point to other points results in a larger value of z as computed according to Equation (38). Therefore the point will belong to the classification where its z value is the highest.

Consider a simple Gaussian pattern equation that computes the result for a point (x_O, x_1) relative to already classified points (x_{Oi}, x_{1i}):

$$z = f(x_0, x_1) = \Sigma_{i=1}^{n} e^{\frac{-\left((x_0 - x_{0i})^2 + (x_1 - x_{1i})^2\right)}{2\sigma^2}} \tag{38}$$

Suppose that you intend to predict the future state of a patient's intra-cranial pressure (ICP). The goal is to classify the input data into one of three possible classes: normal, moderate or critical. Each class is represented by f_N, f_M and f_C that are probability distribution functions for category N, M and C (normal, moderate and critical). Then one can compute f_N, f_M and f_C as follows:

$$z_N = f_N(x_0, x_1) = \Sigma_{i=1}^{n} e^{\frac{-\left((x - x_{0Ni})^2 + (x - x_{1Ni})^2\right)}{2\sigma^2}} \tag{39}$$

$$z_M = f_M(x_0, x_1) = \Sigma_{i=1}^{n} e^{\frac{-\left((x - x_{0Mi})^2 + (x - x_{1Mi})^2\right)}{2\sigma^2}}$$

$$z_C = f_C(x_0, x_1) = \Sigma_{i=1}^{n} e^{\frac{-\left((x - x_{0Ci})^2 + (x - x_{1Ci})^2\right)}{2\sigma^2}}$$

Here the data point to be classified is X and all the other points that belong to other classifications are referred to by N_i, M_i and C_i. There are many activation functions possible but a common non-linear operation is the following:

$$e^{\frac{-(W_i - X)^T (W_i - X)}{2\sigma^2}} \tag{40}$$

If X and W_i are normalized to unit length, it's been demonstrated (Zaruda 1997) that the non-linear operation above can be replaced by (41). The derivation follows by multiplying the numerator terms. That results in:

$$-w_i^2 + 2w_iX - X^2 = -2 + 2w_iX = 2(w_iX - 1)$$

Since the terms w_i^2 and X^2 are normalized to unity, they are replaced by 1. Substituting Z_i for $W_i\,X$ in the above term in the numerator, you can obtain the following activation function:

$$e^{\frac{(Z_i - 1)}{\sigma^2}} \tag{41}$$

So a simple algorithm to identify classification of a new data set can be described as:

1. Input layer: Normalize X and W_i to unit length
2. Pattern layer: Compute the dot product of input X and the weights of X, W_i
3. Summation layer: Compute f_N, f_M and f_C
4. Output (or decision) layer: Select the output with the highest response value (from N, M or C clusters of neurons)

The final classification is determined by a classifier function C that selects the largest of f_N, f_M and f_C values:

$$Prediction_{(t+t1)} = C(f_N, f_M, f_C) = max(f_N, f_M, f_C) \tag{42}$$

An Example in Appendix B illustrates how PNN can be applied to a simple classification problem.

8.12 Support Vector Machine (SVM) Networks

Support Vector Machines (SVM) are among a number of supervised training models that analyze data for multiple classification and regression analysis. SVM is a machine learning method using supervised learning models. It is used to detect patterns and often used for classification & regression analysis. It can handle linear and non-linear data for training the model effectively. The SVM method is now highly regarded for its accuracy of prediction.

The SVM is a non-probabilistic binary linear classifier. It takes a set of input data and determines which of the possible classes the input is a member of. SVM constructs a set of hyperplanes between data elements to classify them.

A good separation is the mark of a generalizable model and is achieved by the hyperplane that has the largest distance to the nearest training data element in any class. The hyperplane is mathematically defined as the set of data elements whose inner product with a vector in that space is constant. *Margin* is the distance between the optimal hyperplane and a vector that runs close to it. The following derivation is adapted from Sengupta (2009) and Haykin (1998), improved and revised specific to this framework.

The most optimum solution can be found by gaining the biggest possible margin. In Figure 8.15, two examples are shown that illustrate the band that separates the data into two categories. This is an example of the binary classification where your goal is to identify which data points fall into one category vs. the other. For example, if you want to predict which customers are likely to leave or remain loyal. In multi-class problems, you're likely to separate data into multiple classifications. The band that separates the data into two or more classifications is called a hyperplane.

The derivation of equations 43–48 are offered to illustrate the math behind hyperplane computation, but can be skipped without loss of information.

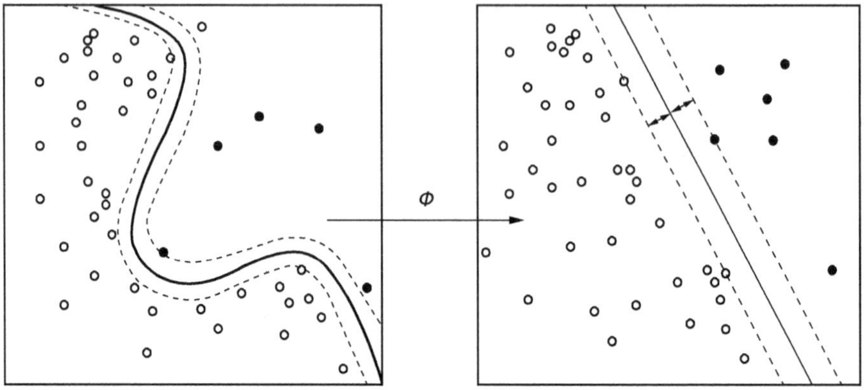

Figure 8.15: Support vector machines as classifiers.

To describe the hyperplane in mathematical terms, the optimal hyperplane must satisfy the following:

$$\frac{y_k F(x_k)}{\|w\|} \geq \tau, \; k = 1, 2, \; \ldots, n \tag{43}$$

where τ is the margin and can be visualized as a band that separates the nearest points from the hyperplane that separates them into two categories (as shown in Figure 8.15). Note that F(x) is defined by:

$$F(x) = w^T x + b \tag{44}$$

One can map the data points to a very high-dimensional space, then the algorithm finds a hyperplane in this space with the largest margin separating classes of data. The feature space is usually defined as a non-linear product of base functions $\varphi_i(x)$, defined in the input space. This non-linear product is explained below. Then function *F(x)* becomes:

$$F(x) = \sum\nolimits_{i=1}^{n} a_i K(x_i, x) + b \tag{45}$$

where $K(x_i, x)$ is the inner product kernel of base functions $\varphi_i(x), j = 1, 2, \; \ldots, \; m$. The cross products in the larger space are defined by a kernel function *K(x, y)* that best fits the problem, such that:

$$\sum\nolimits_i a_i K(x_i, x) = constant \tag{46}$$

It can be observed that *K(x,y)* becomes small as *y* grows further from x. The inner product K(.) can have many possible kernels. Each kernel is essentially a solver that includes a sigmoid function. (Recall from our earlier description of neural networks that the output of each perceptron was shaped by using a sigmoid function). One of the most commonly used is based on the Gaussian:

$$K(x_i, x) = e^{\left(-\frac{\|x - x_i\|^2}{2\sigma^2}\right)} \tag{47}$$

where $\sigma > 0$, and sigmoid kernels $K(x_i, x) = \tanh(\theta < x_i, \ x > +\vartheta)$ such that

$$\theta > 0, \quad \text{and} \quad x > +\vartheta. \tag{48}$$

Figure 8.15 illustrates two lines that separate data into two clusters.

8.13 General Feed-forward Neural Network

A feed-forward neural network consists of one or more layers of nodes where the information flows in only one direction, forward from the input nodes and there are no cycles or loops in the network. The simplest networks have a single layer that feeds input data to the output layer via a series of weights.

One of the popular training methods is the gradient descent algorithm. The following is a mathematical development of the gradient descent method, adapted from Sengupta (Sengupta 2009) and Zurada (Zurada 1997) and improved for this framework.

For a given neural network with *n=1, . . ., N* layers, you can compute the error for each node. By definition the error at time snapshot *n*, for node *j* computed by:

$$e_j(n) = d_j(n) - y_j(n) \tag{49}$$

And for computing total error of a network, recall Equation (21) provides that:

$$E(n) = \frac{1}{2}\sum_j e_j^2(n) \quad \text{for} \quad j = 1 \ to \ m. \tag{50}$$

Then you can compute average error of a network by:

$$E(N)_{Avg} = \frac{1}{N}\sum_{n=1}^{N} E(n) \tag{51}$$

Let's use the traditional Pitts-McCullough equation to compute:

$$v_j(n) = \sum_{i=0}^{m} w_{ji}(n) y_i(n), \tag{52}$$

where $v_j(n)$ is the input to the activation function for the neuron *j*. The term $v_j(n)$ can be perceived as the induced local field, while $y_i(n)$ can be thought of as the output from the previous layer, shown by:

$$y_i(n) = \varphi\left(v_j(n)\right) \tag{53}$$

Given the prior introduction, it's possible to start the mathematical derivation of the back-propagation method. First it's easy to calculate the partial derivative of *E(n)* and apply the chain rule to obtain the following:

$$\frac{\partial E(n)}{\partial w_{ji}(n)} = \frac{\partial E(n)}{\partial e_j(n)} \cdot \frac{\partial e_j(n)}{\partial y_j(n)} \cdot \frac{\partial y_j(n)}{\partial v_j(n)} \cdot \frac{\partial v_j(n)}{\partial w_{ji}(n)} \tag{54}$$

Where Δw_{ij} is applied to w_{ij}. Let's apply the derivatives to each component above (from equation 54):

$$\frac{\partial E(n)}{\partial e_j(n)} = e_j(n)$$

$$\frac{\partial e_j(n)}{\partial y_j(n)} = -1$$

$\frac{\partial y_j(n)}{\partial v_j(n)} = \varphi'(v_j(n))$, and you can take derivative of $\varphi(v_j(n))$ when the exact function is known.

$$\frac{\partial v_j(n)}{\partial w_{ji}(n)} = y_i(n)$$

Now it's possible to write the result of substitutions as:

$$\frac{\partial E(n)}{\partial w_{ji}(n)} = -e_j(n)\varphi'(v_j(n))y_i(n) \tag{55}$$

When adjustment of $\Delta w_{ji}(n)$ (namely the correction to $w_{ji}(n)$ is applied to $w_{ji}(n)$ the following relationship can be obtained:

$$\Delta w_{ji}(n) = -\eta \frac{\partial E(n)}{\partial w_{ji}(n)} = \eta \; e_j(n)\varphi'(v_j(n))y_i(n) \tag{56}$$

This equation is a reminder that $\Delta w_{ji}(n)$ is proportional to $\frac{\partial E(n)}{\partial w_{ji}(n)}$ at a rate of proportionality η, negative to the direction of the gradient. Part of the term in (55) can be described as error multiplied by the activation function. This term can be shown as:

$$\partial_j(n) = -\frac{\partial E(n)}{\partial v_j(n)} = -\frac{\partial E(n)}{\partial e_j(n)} \cdot \frac{\partial e_j(n)}{\partial y_j(n)} \cdot \frac{\partial y_j(n)}{\partial v_j(n)} \tag{57}$$

$$\partial_j(n) = e_j(n)\varphi'(v_j(n)) \tag{58}$$

The term $\partial_j(n)$ is the derivative of error with respect to the activation function $v_j(n)$. This is the gradient for neuron *j* and is local to neuron *j*. This is also known the *local gradient*. Now one can rewrite Equation (56) as:

$$\Delta w_{ji}(n) = \eta\, \partial_j(n) y_i(n) \text{ where } y_i(n) \text{ is the input to neuron } j. \tag{59}$$

Since the values for η and $y_i(n)$ are known it's possible to compute $\partial_j(n)$. Two cases are possible:

Case 1) Neuron j belongs to the output layer, hence $\partial_j(n)$ can be calculated from Equation (54).

Case 2) Neuron j belongs to a hidden layer, thus $\partial_j(n)$ must be computed differently.

A fundamental condition for back-propagation is that you can compute $\partial_j(n)$, namely that the derivative of the activation function is possible. Let's examine this approach graphically to illustrate the hidden layer and output layer computations as shown in Figure 8.16. The signal flow graph in Figure 8.16 shows that j is the hidden layer and k is output layer neurons.

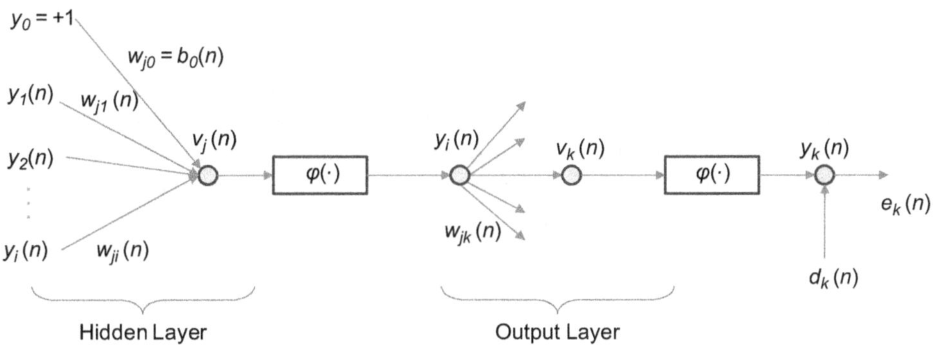

Figure 8.16: Depiction of hidden layer j and output layer k neurons.

You can compute $\partial_j(n)$ by the following derivations. It's known that:

$$\partial_j(n) = -\frac{\partial E(n)}{\partial v_j(n)} = -\frac{\partial E(n)}{\partial y_j(n)} \cdot \varphi'(v_j(n)) \tag{60}$$

Also recall from Equation (50) that:

$E(n) = \frac{1}{2}\Sigma_k e_k^2(n)$, where k is the set of output layer neurons. It can be shown that:

$$\frac{\partial E(n)}{\partial y_j(n)} = \Sigma_k e_k(n)\frac{\partial e_k(n)}{\partial y_j(n)} = \Sigma_k e_k(n)\frac{\partial e_k(n)}{\partial v_k(n)} \cdot \frac{\partial v_k(n)}{\partial y_j(n)} \tag{61}$$

$$\text{Since } e_j(n) = d_j(n) - y_j(n) = d_j(n) - \varphi(v_j(n)) \tag{62}$$

Let's take derivatives of the left hand side:

$$\frac{\partial e_k(n)}{\partial v_k(n)} = -\varphi'(v_k(n)) \tag{63}$$

For neuron k, it's possible to write:

$$v_k(n) = \Sigma_{j=0}^{m} w_{kj}(n) y_j(n) \tag{64}$$

Consequently by taking the derivative of (64), it results in the following:

$$\frac{\partial v_k(n)}{\partial y_i(n)} = w_{kj}(n) \tag{65}$$

$$\frac{\partial E(n)}{\partial y_j(n)} = -\Sigma_k e_k(n)\varphi'(v_k(n)) w_{kj}(n) = -\Sigma_k \partial_j(n) w_{kj}(n) \tag{66}$$

From Equation (60) and (66) one can conclude that:

$$\partial_j(n) = \varphi'_j(v_j(n)) \cdot \Sigma_k \partial_k(n) w_{kj}(n) \tag{67}$$

This equation is significant as it shows that the local gradient of neuron j (hidden neurons) depends on the local gradient of output neuron k. The summation is the weighted sum of all the output gradients. Let's assume M is the number of output neurons. Consider multiplying each error term $e_j(n)$ by the derivative of the corresponding activation function, namely $\varphi_k(v_k(n))$ by $\varphi'_k(v_k(n))$. This is the basis of back-propagation as depicted in Figure 8.17.

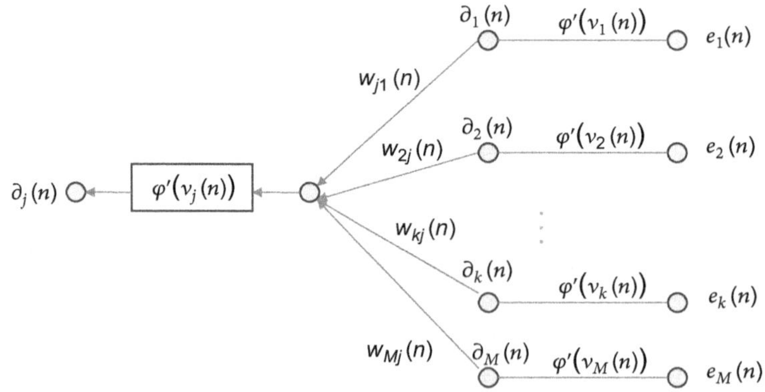

Figure 8.17: Backward propagation—computing $\partial_j(n)$ from errors $e_k(n)$ in forward step.

Once a backward propagation is completed, the process flow reverses by taking a forward pass. The forward pass re-computes the results of the neural net with the new weights. In the forward pass, the inputs propagate forward from the first layer to the second layer and so on, eventually to the output layer. The error terms are computed and then the backward propagation begins. In backward propagation, the algorithm starts with the error term. Then it calculates the local gradients and propagates back and adjusts the synaptic weights.

The activation function can be any of the three types mentioned previously, such as the sigmoid function, logistic function or a *tanh(.)* function. Let's compute the derivative of the function for the case of the logistic function:

$$\varphi_j(v_j(n)) = \frac{1}{1+e\left(-av_j(n)\right)} \equiv y_j(n) \tag{68}$$

It's given that $\alpha > 0$, and $-\infty < v_j(n) < \infty$. It's possible to compute the derivative of the activation function as follows:

$$\varphi'_j(v_j(n)) = \frac{a \cdot e^{(-av_j(n))}}{[1+e^{(-av_j(n))}]^2} = a \cdot y_j(n)[1-y_j(n)] \tag{69}$$

In other words, you can write the above in the following fashion:

$$\varphi'_j(v_j(n)) = [1-y_j(n)] \cdot (a) \cdot (y_j(n)) \tag{70}$$

Now you can use $y_j(n)$ for computing $\partial_j(n)$ by:

$$\begin{aligned} \partial_j(n) = \varphi'_j(v_j(n)) \cdot \sum{}_k \partial_k(n) w_{kj}(n) = \\ a \cdot y_j(n)[1-y_j(n)] \sum{}_k \partial_k(n) w_{kj}(n) \end{aligned} \tag{71}$$

The range of $y_j(n)$ is [0,1]. The value of $\varphi'_j(v_j(n))$ is maximum when $y_j = 0.50$, and $\varphi'_j(v_j(n))$ is equal to zero when $y_j = 1.0$ or zero. This fact guides our choice of proper values for Δw_{kj}. Similarly, you could have taken the derivative of $tanh(x)$, to compute Δw_{kj}. A detailed description of this algorithm appears in Appendix C.

8.14 MLP with Levenberg-Marquardt (LM) Algorithm

Feed-forward MLP with LM is a feed-forward neural network that consists of one or more layers of nodes where the information flows in only one direction, forward from the input nodes and there are no cycles or loops in the network (Sengupta 2009, Masters 1995). The simplest networks have a single layer that feeds input data to the output layer via a series of weights. In the multi-layer perceptron model (MLP), each node has direct connection to the nodes in the subsequent layer.

The sum of products of the weights and the inputs are calculated in each node (Haykin 1998). If the value of the result is above a certain threshold, the neuron fires with the activated value (typically 1), otherwise, it fires the deactivated value (typically -1). Several variations of this model and training methods have been proposed, including the backward propagation algorithm and Levenberg-Marquardt (LM) method which is considered as one of the more computationally efficient algorithms. The following derivation is adapted from Sengupta (2009), Haykin (1998) and Masters (1995), revised and improved for this framework.

The training of MLP occurs in two stages: in a forward phase the weights of the network are fixed and the input data is propagated through the network. The forward phase completes its computation with an error signal. The error term was defined by Equation (49) and can be defined as

$$e_{kp} = d_{kp} - y_{kp},\ k = 1,\ \ldots,\ K,\ p = 1,\ \ldots,\ P \tag{72}$$

where d_{kp} is the desired response and y_{kp} is the actual output produced by the network response to the input x_{ip}. d_{kp} is the desired value of the kth output and the Pth layer. Y_{kp} is the actual value of the kth output and P^{th} pattern. The parameter K is the number of network outputs, P is the number of patterns and N is the number of weights.

In the backward phase, the error e_{kp} is propagated through the network going backward and the free weights are adjusted to minimize error e_{kp}. In the LM algorithm, the performance index *F(W)* is to be optimized:

$$F(W) = \sum_{p=1}^{P} \left[\sum_{k=1}^{K} (d_{kp} - y_{kp})^2 \right] \tag{73}$$

Where W = $[w_1\ w_2 \ldots w_N]^T$ is the set of all weights for the network. The equation can be written as:

$$F(W) = E^T E$$

Where

$$E = [e_{11} \ldots e_{K1}\ e_{12} \ldots e_{K2} \ldots e_{1P} \ldots e_{KP}]^T \tag{74}$$

The error term E, is the cumulative error vector for all patterns. Let's assume that the amount of change to each weight is shown by θ_{ij}. Using the Jacobian matrix, you can compute the amount of change that you want to be applied to weights in the backward propagation. By definition, a Jacobian is the derivative of one vector with respect to another vector. In vector calculus, the Jacobian matrix is the matric of all first order partial derivatives of a vector-valued function. From the equation above, the Jacobian matrix can be defined as:

$$J = \begin{bmatrix} \frac{\partial e_{11}}{\partial w_1} & \frac{\partial e_{11}}{\partial w_2} & \cdots \frac{\partial e_{11}}{\partial w_N} \\ \frac{\partial e_{21}}{\partial w_1} & \frac{\partial e_{21}}{\partial w_2} & \cdots \frac{\partial e_{21}}{\partial w_N} \\ \vdots & \vdots & \vdots \\ \frac{\partial eK_1}{\partial w_1} & \frac{\partial eK_1}{\partial w_1} & \frac{\partial eK_1}{\partial w_1} \\ \vdots & \vdots & \vdots \\ \frac{\partial e_1P}{\partial w_1} & \frac{\partial e_1P}{\partial w_2} & \cdots \frac{\partial e_1P}{\partial w_N} \\ \frac{\partial e_2P}{\partial w_1} & \frac{\partial e_2P}{\partial w_2} & \frac{\partial e_2P}{\partial w_N} \\ \vdots & \vdots & \vdots \\ \frac{\partial eKP}{\partial w_1} & \frac{\partial eKP}{\partial w_2} & \frac{\partial eKP}{\partial w_N} \end{bmatrix} \tag{75}$$

Using the Newton-Raphson method, you can compute the change in weight by applying a dampened measure of error terms. We'll use the Jacobian matrix (shown as J in the following formula) to compute the change to Levenberg-Marquardt model. It's possible to derive the Levenberg-Marquardt algorithm as follows: Recall that the output is a function of inputs x_i and weights W. This can be stated by writing:

$$y_i = y_i(x_i, W)$$

In other words, y_i depends on input x_i and weights W. The goal of backward propagation is to adjust the weights W by $J_i\delta$ where δ is the amount of adjustment and J_i is the ith row of the Jacobian matrix. Then it's possible to write this expression as:

$$y_i(x_i, W + \delta) \cong y_i(x_i, W) + J_i\delta$$

The goal of computation is to minimize the objective function in Equation (73):

$$\text{Minimize } E(W) = \sum_p \sum_k (d_i - y_i)^2$$

The variable δ, represents the small change or adjustment to the weight. When adjusted by δ, the minimization function can be written as:

$$E(W + \delta) = \sum_p \sum_k (d_i - y_i(x_i, W + \delta))^2 \cong \sum_p \sum_k (d_i - y_i - J_i\delta)^2$$

Next the error term can be written as:

$$E(W + \delta) \cong \left\|\bar{y} - \bar{d} - J\delta\right\|^2 \tag{76}$$

To minimize this function, you take the derivative and sets it to zero. The derivative of the above function set to zero becomes:

$-2J^T(\bar{d}-\bar{y}-J\delta)=0$, namely:

$$J^T(\bar{y}-\bar{d})=J^TJ\delta$$

The rate of change can be tempered by a scalar multiple shown by parameter η. Equation (76) can be written as:

$J^T(\bar{y}-\bar{d})=(J^TJ+\eta I)\delta$, or as:

$$J^TE=(J^TJ+\eta I)\delta$$

where *I* is the identity matrix. It can be seen that the rate of change in weights can be computed by:

$$\delta=\frac{J^TE}{(J^TJ+\eta I)}$$

It can be easily seen that the weights for the next iteration can be computed using the following equation:

$$W_{t+1}=W_t-(J_t^TJ_t+\eta_tI)^{-1}J_t^TE_t \tag{77}$$

Where ***J*** is the Jacobian (a matrix of first order derivatives) of *m* input errors with respect to *n* weights of the neural network, ***I*** is the identity matrix and η_t is the learning parameter.

Neural networks offer a variety of structures and learning methods that deliver a great deal of flexibility and power for a wide range of applications. Programming environments such as Python Scikit-Learn, Tensorflow and R provide various neural network and deep learning programs that support many of these structures.

A key aspect of modeling with neural networks is accuracy and model performance. In the next chapter, we'll review how to measure model performance and improve its accuracy.

Chapter 9
Model Accuracy and Optimization

The traditional data-driven prediction methods have constructed multiple models and selected the one with the best performance, discarding all the others. This approach has some disadvantages:

1) The effort of constructing several models is wasted.
2) The selected model may not consistently perform most accurately or be robust on all types of data and with diverse types of predictions.
3) The selected algorithm may not be able to sustain or perform consistently on training data types as changes to data types occur over time.

To overcome these disadvantages, I've effectively applied a multi-model ensemble approach (also known as committee of models) which combines multiple algorithms with a weighted sum formulation.

Model performance is defined by many attributes including model accuracy and computational speed. I'd like to focus on accuracy as a key factor to improve in this session. Model optimization here means how to maximize a model's accuracy. I introduce this framework with five different ensemble schemes and compare their performance on the medical prediction case study data set.

These schemes include a voting formula, two accuracy-based weighting schemes, a diversity-based weighting and optimization-based weighting. The goal of constructing ensembles is to identify the weights of each algorithm such that it improves data-driven prognostics performance.

The case study employed in this book demonstrates that the ensemble approach provides a more accurate prediction than any single best algorithm. Given a number of neural networks to select from, the goal is to select a weighted sum of these models' output that provides the most accurate classification.

An oracle program can be defined to select the most accurate algorithm from a set of five ensembles (or committee) of algorithms provided by the four ANN models. The oracle is defined as an overseer program which selects the most appropriate answer among a set of options. The oracle, as an overseer selects a prediction from among a number of ensembles or models that meet a desired level of accuracy or prediction characteristic. The performance of each model is determined based on its accuracy on the validation step using the test data set, also known as the set-aside data set.

Since one model may perform better in predicting true positives and another better at predicting the true negatives on a particular data set, the oracle program combines the predictions from multiple models in a way that the model with higher accuracy is assigned a higher weight than the worse model. Even the worse model

https://doi.org/10.1515/9781547401567-010

contributes to the prediction but at a smaller weight. Training on a different data set might be completely the reverse. Hence, the ensemble adjusts its proportion of each model to deliver the highest accuracy desired. In this way, the oracle can improve the classification accuracy, sensitivity and specificity by combining the best classification characteristics from different models.

Given that there are many neural networks to select from, the goal is to select the most accurate model, or ensemble of models to predict if a patient is going to contract a particular disease. Let's as an example suppose that the prognostics engine trains four different algorithms to make a prediction for DVT. It can then build five different ensembles with different weighted sums from the four models' output.

Consider for example a study to classify patients using machine learning classifiers (using their test data) into two groups: those who will be sick and those who will be healthy. So, a positive case is a patient with illness and a TP classification means the algorithm has correctly classified the ill patient as sick. Then FP represents a Type I error (α) and FN represents a Type II error (β) as shown in Figure 9.1.

<table>
<tr><td></td><td colspan="3">Truth</td></tr>
<tr><td rowspan="3">Test</td><td></td><td>Sick</td><td>Healthy</td></tr>
<tr><td>Sick</td><td>TP
(1-β)</td><td>FP
α</td></tr>
<tr><td>Healthy</td><td>FN
β</td><td>TN
(1-α)</td></tr>
</table>

Figure 9.1: Truth Table indicating Type I and Type II errors.

The oracle program can be set up to enhance any of the four desired accuracy characteristics: either directly improve true positives (TP), true negatives (TN), false positives (FP) or false negatives (FN), or to improve other accuracy measurements such as sensitivity, specificity or Youden's *J* index.

It's important to define the objectives for the oracle program. The practitioner must define the selection criteria for the oracle program; is the oracle to select a model or ensemble of models that reduce Type I (FP) error, or Type II (FN) error, or both errors; or as it will be presented later, a combination of accuracy measures.

While in most situations, the criteria call for lower Type II error, in certain situations where the population is large and the cost of treatment is high, the criteria would include reducing Type I error as well.

The oracle program can be set up to provide a meta-classification based on a linear combination of these characteristics. The goal of the ensemble is to produce a synthesized classification result based on weighted sum of ANN models as:

$$\hat{p} = \sum_{j=1}^{K} (w_j p_j)$$

The assumption in this book is that the ensemble consists of K models, and p_j is the prediction of model j. The result of the ensemble's prediction is represented by $\hat{p}$.

9.1 Accuracy Measures

The accuracy measures may be defined as:

$$accuracy = \frac{TP + TN}{TP + TN + FP + FN}$$

$$sensitivity = \frac{TP}{TP + FN}$$

$$specificity = \frac{TN}{TN + FP}$$

$$LR+ = \frac{sensitivity}{1 - specificity} = \frac{Pr(T+|D+)}{Pr(T+|D-)}$$

$$LR- = \frac{1 - sensitivity}{specificity} = \frac{Pr(T-|D+)}{Pr(T-|D-)}$$

The *likelihood ratio* combines both sensitivity and specificity into a single measure. It provides a direct estimate of how much a test result will change the odds of having a disease.

The Positive LR (LR+) shows how much the odds of the disease increase when a test is positive. The Negative LR (LR-) shows how much the odds of the disease decrease when a test is negative. Odds can be derived from probability. To convert from probability to odds, divide the probability by one minus that probability. Positive LR is the ratio of sensitivity to one minus specificity (Delen 2009).

9.2 Accuracy and Validation

External validity of medical prediction models is an extremely challenging task. Clinical validation is challenging not just because it involves prospective patient studies, double-blind studies and careful administration of research protocols, but for two other reasons: first, if the model predicts a disease and the patient gets the treatment per recommendation of the predictive model, we can't determine if the patient would have exhibited the predicted disease to confirm our prediction. In other words, the medical treatment masks the possible outcome. Second and in contrary, if the model predicts no disease but the patient gets treatment, we would not be able to invalidate the model's prediction since we can't claim that the disease might have occurred.

This framework focuses on internal validity in terms of accuracy but leaves external (real life) validation to future research projects. Several measurements have been proposed as methods for internal validation. Some of the measurements that are commonly used to compare accuracy of classification models include: *accuracy*, *sensitivity*, *specificity*, *area under receiver operating curve* (AUROC) and *likelihood ratio* (LR). *Sensitivity* measures the fraction of positive cases that are classified correctly as positive.

Specificity is the fraction of negative cases that are classified correctly as negative. AUROC is the area under the ROC and is regarded as a good overall measure of predictive accuracy of a model (Bewick, Cheek, Ball 2004). A ROC can be plotted by connecting the points obtained from ANN model results at different model thresholds as shown in Figure 9.2.

A ROC is a graph that represents a plot of *sensitivity* versus *(1—specificity)*. The Area under the ROC curve (AUROC) can be computed by the sum of trapeziums[58] areas under the curve. An *AUROC* close to 1.0 is a considered excellent discrimination, but a value near 0.50 suggests no discrimination (similar to a coin flip).

The ROC curve for each model was computed and compared as shown in Figure 9.2. From a visual inspection, it's clear to see that the SVM model has more desirable accuracy due to its larger relative area under the ROC (AUROC).

The likelihood ratio combines both sensitivity and specificity into a single measure. It provides the direct estimate of how much a test result will change the odds of having a disease. The Positive LR (LR+) shows how much the odds of the disease increase when a test is positive.

The Negative LR (LR-) shows how much the odds of the disease decrease when a test is negative. Odds can be derived from probability. As mentioned before, convert from probability to odds, divide the probability by one minus that probability.

Positive LR is the ratio of sensitivity to one minus specificity (Delen 2009). An LR+ is a ratio, equal to the probability of a person who has the disease and tested positive divided by the probability of a person who does not have the disease and tested positive. An LR- is another ratio, equal to the probability of a person who has the disease and tested negative divided by the probability of a person who does not have the disease and tested negative.

The likelihood ratio is useful when the pre-test odds of having a disease are known. Then, the post-test odds of disease can be computed by:

58 Divide the area under the curve into vertical slices by adding some straight vertical lines. Each slice is a trapezium.

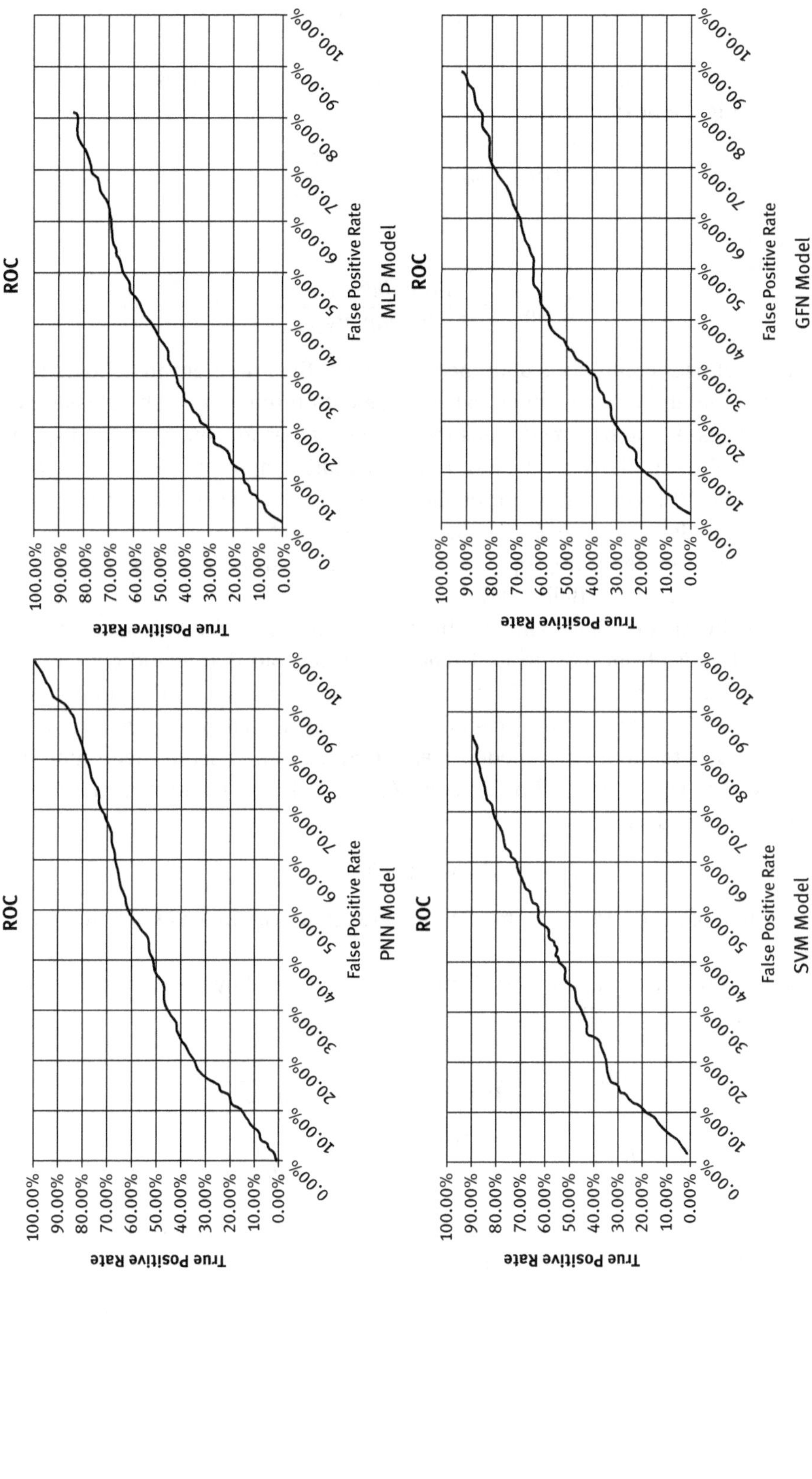

Figure 9.2: ROC Curves for the four ANN models.

$$odds_{post-test} = odds_{pretest}{}^{*}likelihood\ ratio$$

This calculation is based on Bayes Theorem. You can convert odds to probability simply by using the next formula.

$$Probability_{pretest} = \frac{TP + FN}{Total\ Sample}$$

Alternatively,

$$Odds_{pretest} = \frac{Probability_{pretest}}{1 - Probability_{pretest}}$$

When a model uses continuous data measurements, then different thresholds may be applied in order to decide which value is the cut-off to distinguish between patients with disease. The best model has the highest values for sensitivity and specificity. In certain situations, both may not be equally important. For example, a false-negative (FN) prediction might be more critical than a false-positive (FP) prediction. If no preference is given to either measurement then, Youden's index (*J*) may be used to choose an appropriate cut-off, computed by using specificity and sensitivity (Bewick, Cheek, Ball 2004).

When using ANN models to make a binary prediction, the result is a continuous measure that varies from 0.00 to 1.00. The ideal threshold that would classify patients into disease or healthy can be set using Youden's index *J*. At the point where Youden's index is highest, the threshold can be set at that level. The Youden's index for each of the four ANN models were computed and used to determine the ideal threshold. The relationship between Youden's index *J* and sensitivity and specificity is defined as:

$$J = sensitivity + specificity - 1$$

A higher value of *J* is desired. The maximum value that *J* can take is 1, when the test is perfect.

PPV corresponds to the number of true positives divided by the sum of true positives and false positives. NPV is computed as the ratio of

$$PPV = \frac{True\ Positives}{(True\ Positives + False\ Positives)}$$

$$NPV = \frac{True\ Negatives}{(True\ Negatives + False\ Negatives)}$$

Figure 9.3 shows a statistical truth table (Also known as a Confusion matrix) that illustrates how sensitivity, specificity, PPV and NPV are related. In this figure, the results of the GFN model are used only as an illustration to show how PPV, NPV, Sensitivity and Specificity are calculated.

		Truth (Condition) as determined by "Gold Standard"		
		Positive Condition	**Negative Condition**	
Test Outcome	Test Outcome: **Positive**	**True Positive** (TP) = 129 1−β	**False Positive** (FP) = 336 (Type I error) A	Positive Predictive value = $\frac{\sum \textit{True Positive}}{\sum \textit{Test Outcome Positive}}$ = TP/(TP+FP) =129/(129+336) =**27.7%**
	Test Outcome: **Negative**	**False Negative** (FN) = 96 (Type II error) β	**True Negative** (TN) = 512 1−α	Negative Predictive value = $\frac{\sum \textit{True Negative}}{\sum \textit{Test Outcome Negative}}$ = TN/(FN+TN) = 512/(96+512) = **84.2%**
		Sensitivity $\frac{\sum \textit{True Positive}}{\sum \textit{Condition Positive}}$ = TP/(TP+FN) = 129/(129+96) = **57.3%**	**Specificity** $\frac{\sum \textit{True Negative}}{\sum \textit{Condition Negative}}$ = TN/(FP+TN) = 512/(336+512) = **60.4%**	

Figure 9.3: Statistics Truth Table (Confusion Matrix).

In statistics and medicine, the gold standard test refers to a diagnostic test that is the best available to diagnose or classify a patient into either a disease or normal condition. The gold standard test is not necessarily a perfect test, but one that offers the most accurate test possible without restrictions.

The ideal gold standard test offers 100% sensitivity and specificity. In practice, however, a gold standard test is less accurate. For example, to diagnose a brain tumor, you can perform a biopsy or an MRI. A biopsy test is regarded as the gold standard for diagnosing a brain tumor, but since the MRI test is less accurate but a practical substitute, it's regarded as an "imperfect gold standard" or an "alloyed gold standard" (Spiegelman, Schneeweiss, McDermott 1996). Gold standard tests vary over time for each disease as the state-of-the art methods of diagnostic tests improve over time.

9.3 Vote-based Schema

A *voting approach* selects the final results based on majority vote on a specific accuracy dimension or result category. For example, if the goal is to minimize FN, then you must consider a voting schema that produces the lowest FN among the four models on a given new case data set. The voting schema works as follows: Run all four models on the new data set. Select the lowest FN count, or select a weighted sum of models that produce the lowest FN+FP count.

A hybrid schema can work by defining a relative preference ratio. The assumption is that models with more correct predictions are preferred over the other models. The relative preference of a model is determined by the number of correct predictions minus incorrect predictions. This schema was used for Ensemble #1. The mathematical representation of schema for Ensemble#1 is shown in (79) and (80). This ensemble assumes n data cases and K models in the ensemble:

$$model\ j\ score = \sum_{i=1}^{n} (correct\ predictions - incorrect\ predictions) \tag{79}$$

$$W_j = \frac{model\ j\ score}{\sum_{i=1}^{K} model\ i\ score} \tag{80}$$

As shown in this equation, the intent is to find the weights according to the ratio of preference for a model that provides higher accuracy (lower FN count).

9.4 Accuracy-based Ensemble Schema

An accuracy-based ensemble seeks to find a weighted sum of algorithms that minimize error. For example, if the intent is to find the lowest FN, you can define the error for each validation case as follows. Let's assume that T_i is the truth of a particular validation case, and p_{ij} is the prediction rendered by algorithm j for case i. Also assume that there are a total of K different algorithms. Ensemble #2, attempts to increase the number of the TP count by defining the number of errors for each model j:

Then you can define the total error count by:

$$\varepsilon_j = \sum_{i=1}^{n} \varepsilon(p_{ij}, T_i)$$

where $\varepsilon(p_{ij}, T_i)$ equals to 1 if p_{ij} does not match the truth T_i, or equals zero if prediction matches the truth.

Then you can define the weight of algorithm j for ensemble #1 with the goal of increasing the quantity of TP. Assuming that the total number of TP from a model j is shown by TP_j, it's possible to define the weights by:

$$W_j = \frac{(TP_j)}{\sum_{j=1}^{m}(TP_j)}$$

For ensemble#3, you can define the weight of algorithm j with the goal of reducing the quantity of FN error:

$$W_j = \frac{(\varepsilon_j)^{-1}}{\sum_{j=1}^{K}(\varepsilon_j)^{-1}}$$

In this framework, two ensembles, Ensemble#2 and Ensemble#3 use an accuracy-based scheme to determine the weights. The results are discussed and compared in Chapter 5.

9.5 Diversity-based Schema

The accuracy-based weighted sum formulation exclusively relies on accuracy to compute the weights. However, it could be argued that accuracy is not the only factor that affects ensemble performance. The diversity-based schema measures the extent to which the predictions by one model are distinguishable from predictions by other models. The diversity-based schema increases the robustness performance of the ensemble. Said differently, this ensemble assigns higher weight to the model with higher prediction diversity because it offers higher ensemble robustness. Let's assume that n data sets are employed to train and validate all k models. You can compute a prediction error term u_j to correspond to the uniqueness of the model j:

$$\theta_j = (p_{j1} - T_1, \ldots, p_{jn} - T_n)$$

Given k algorithms, it's possible to define the error vector of $\theta_1, \theta_2, ..., \theta_K$. The diversity of the jth model can be computed as the sum of the Euclidean distances between the vector θ_j and all other error vectors, defined by:

$$D_j = \sum_{i=1; j \neq i}^{K} \theta_j - \theta_i$$

The prediction diversity determines how a model's result is distinguishable from those of other models. Let's compute a normalized weight w_j for the jth model in the ensemble by:

$$W_j = \frac{D_j}{\sum_{i=1}^{K} D_j}$$

Ensemble#4 uses diversity-based schema.

9.6 Optimization-based Schema

One proposal in this book is to define the oracle (meta-classifier) as an optimization model. The optimization-based schema can take into account both the accuracy-based as well as the diversity-based weighting scheme to improve accuracy and robustness. This method is adapted from Hu et al. (Hu, Youn, Wang 2010). In an optimization-based schema you can write Equation (78) as:

$$minimize\ \varepsilon\left(\sum_{j=1}^{K}(w_j p_{ij}), T_i\right) \tag{78}$$

$$\sum_{j=1}^{K} W_j = 1., \text{ and } W_j \geq 0 \ for\ all\ j = 1, \ldots K$$

Where w_j is the weight of model p_j and T_i is the expected result for the ith data set. The objective function attempts to minimize the difference between the expected result and the weighted sum of each model's result. Ensemble#5 uses an optimization-based schema. The weights w_1, w_2, w_3 and w_4 corresponding to each of the four analytical models can be determined according to this optimization-based schema.

Part III: Case Study—Prediction and Advanced Analytics in Practice

Chapter 10
Ensemble of Models—Medical Prediction Case Study: Data Types, Data Requirements and Data Pre-Processing

In our case study, the application of prognostics health management (PHM) to human physiological measurement data promises to conceptually deliver several benefits such as:

1. By continuously and iteratively assessing the physiological and biological input data, it can provide predictions and advanced warning of medical complications.
2. A localized prediction tool that incorporates nuances of local populations and clinical environments.
3. Ability to analyze large data sets and provide trained models for prediction.

Continuous and periodic monitoring of key metrics and KPI's involves collecting historical data from a wide range of internal and external business systems, sensors and streams. The PHM method considers prognostic models built upon prior data from retrospective cases and current data streams in order to make predictions.

Researchers have studied machine learning, classification, adaptive learning methods, and have identified two approaches to machine learning: *programmed* and *concept attainment*. Programmed learning is applicable when the researcher knows the underlying causal relationships in the system and the data.

Concept attainment is an adaptive learning approach where a system learns from an *a priori* set of examples and thus retains concepts from prior data sets. In other words, it learns by classifying patterns in the input data. A concept can be described as a mapping between an input data set and a clinical outcome.

Some interesting research questions arise that merit a full investigation elsewhere, but are discussed briefly here:

A) How much data is adequate for learning to predict a given disease?
B) How much of the error in prediction can be attributed to noise in the data versus flaws in the model that captures the concept or due to the change in concept?
C) What pre-processing methods are applied to measured data to prepare data for learning algorithms?

The answers are briefly discussed as follows.

https://doi.org/10.1515/9781547401567-011

10.1 How Much Data Is Needed for Machine Learning?

In order to train a predictive engine adequately, the ideal data set must consist of an adequate number of both positive and negative cases of prediction (Principe, Euliano, Lefebvre 1999). From experience, some basic rules of thumb have been developed for determining the amount of data needed for proper training of ANN models. The rules have been developed as guidelines by ANN experts (Principe 2011).

1. A data set must have a minimum of five times as many exemplars (data sets) as the number of weights in the ANN model. For example, in the case study in this book which includes 20 input variables, and four layers of network, there can be as many as 20(4-1) or 60 weights. Thus the minimum number of input data should be somewhere around 300 rows of data.
2. The data set must have approximately 50 times the number of exemplars (data samples) than features (independent variables). Exemplars are rows in the data set and features are columns in the data set and. In the case study in this book, I included 20 columns, so it's recommended to have a minimum of 1000 rows of data (50 times more rows than columns) for training the model.
3. In classification models, the number of rows in the smallest class of data should be at least 5 times the number of input columns. For example, in the case study in this book, we are dealing with only two classifications (patient has the onset of disease or no-disease), so the number of data rows in the smaller class must be over 100 rows (5 times 20 input variables).

In order to properly train a model, it's recommended to include an equal number of positive and negative cases to have a balanced network. Let's suppose the goal is to train a model with 400 negative cases and only 50 positive cases, the ANN model naturally works to minimize errors. As a result, the model gets trained by the input data such that minimizing errors for the 400 cases overshadows the minimizing errors for the 50 positive cases. So, it's recommended to randomly select 50 cases from the negative cases and 50 cases from positive sample and train on this set.

10.2 Learning Despite Noisy Data

As described before, one machine learning technique is for the model to learn from the underlying relationships in data in an unsupervised situation. The model learns the concept of the data by training on the dataset. Think of *concept* as patterns in the dataset. We want to train the model that best captures the patterns (or concept) of the dataset.

How much of the error in prediction can be attributed to noise in data versus the model that captures the concept or due to the change in concept? Data noise can be caused by incorrect measurements of the data, disruptions to measurements (missing data) and or incorrect *a priori* classification or units of measure (for example, measuring kilometers but entering it as miles).

Concept issues on the other hand can be attributed to changes to how subjects in a study, or protocols and practices have changed over time. For example, certain diseases develop resistance to certain drugs over time so those drugs are not as effective, or the procedures for a treatment change that make it either more or less effective, but they are still prescribed by same names. These changes contribute to noisy data and errors in prediction.

By conducting a comparison of results of multiple models in an ensemble, the difference in accuracy among the models can give us some clues as to the amount of error that can be attributed to the model.

10.3 Pre-processing and Data Scaling

It's important to study the data before training a model. If the input data ranges are vastly different (for example, one data column has a range of 1000 and another has a range of 1), then the errors from multiplying weights by inputs that have a large range will overtake the smaller data range.

What pre-processing tools are used to prepare measured data for learning algorithms? In most neural network models, all data in each column is normalized such that an amplitude and an offset are applied to the data in each column.

The goal of pre-processing is to bring the range of input data values within a computationally acceptable range for the ANN computations. ANNs train faster and perform better if their data is pre-processed.

The same pre-processing must be applied to test data. Data scaling is an important pre-processing step before training ANNs. Data scaling equalizes the importance of input variables at the input layer. For example, if one input variable ranges between 1 and 10,000 and another ranges between 0.001 and 0.1, the modeler should use proper initial weights, namely small weights for the first variable but larger weights for the second variable.

Data scaling will make the choice of initial weights for ANNs easier so they can train faster. At the input layer, several methods are available to pre-process the measured data for ANN models. These methods include:

1. **Moving average:** Computes the moving average of a column using the chosen window length (the window is the number of selected values for calculation).
2. **Difference:** Computes the difference or percent difference along a column of data from the mean of the column.
3. **Clip data:** Clip data to a given max value or min value (namely, throw out values greater or less than a threshold).

4. **Log of data:** Takes the logarithm of each data item. Log of values is often taken where the variable spans several orders of magnitude such as income. Since the vast majority of incomes are small and a few are very big, it makes sense to scale the data by taking the Log of those variables.
5. **Mean and Variance:** Normalize the data by fixing the mean to zero and use variance from the mean to scale the data.

A commonly used scaling method normalizes input data to unit length. Normalizing to unit length implies that the sum of squares of values in a given data set must equal to 1. To normalize each data value, the following steps are taken:
1. Square all data values in a given data set
2. Sum the squares.
3. Take the square root of the sum of squares.
4. Divide each data item in the data set by this square root of the sum of squares. The result of each division is a normalized data item.

Other pre-processing and coding techniques are necessary for categorical data. For example, let's assume a particular patient's data is recorded as a categorical input variable such that it takes values of Very Low, Low, Medium, High and Very High. This variable should not be coded in numeric values of 0.0, 0.25, 0.50, 0.75, 1.00, as this would create incorrect interpretations by the ANN model.

For example, it would incorrectly imply that a high category is exactly 3 times more than a low value. Such input variables are coded and transformed into binary inputs. In this example, the categorical values would be mapped as shown in Table 10.1.

Table 10.1: Coding example of categorical data.

Category	Binary Input
Very Low	0 0 0 0
Low	1 0 0 0
Medium	1 1 0 0
High	1 1 1 0
Very High	1 1 1 1

Additional data pre-processing occurs at each layer of ANN models. After each layer, the range of normalized values is determined by the range of outputs of the nonlinear transfer functions (activation functions) used by the model. For example, if a *Tanh* axon is used, the data output is normalized to a range between −1 and +1.

If the data for the column is already in the desired range, then the amplitude will be 1 and the offset will be 0, so that the normalized data will be the same as the original data. There are other options that clean and randomize data depending on what is intended for the initial data processing. Some example methods are:

- **Randomize rows:** Randomly re-arrange rows of data.
- **Clean data:** Replace missing or corrupted data with an average of the column, or the most recurring data or the nearest value in the column.
- **Feature selection:** Classifiers define a decision boundary between two classes of data. Features transform input data into a form where classifiers can learn the decision boundary that classifies data. Feature engineering is a difficult task. Some apply a two-phase approach as described below.

Generally, neural networks are less susceptible to missing data or noisy data. One approach to feature engineering is to apply machine learning (neural network models) twice. The first pass builds a general and broad model encompassing all input variables to identify which features (independent input variables) contribute most significantly to prediction and training. In the second attempt, you train the model on these selected features. The result is that your model will train faster and will be more accurate.

There are other techniques for cleansing data; some of those were already discussed in Chapter 6.

The input data may come from a number of sources: acquisition devices, streaming websites, web crawlers and databases.

Once the data is collected into a single database, it can be used to train several classifiers that can classify a new data set into a number of categories.

For the case study, I trained four machine learning models using the N-leave-out method. In the N-leave-out method, each model is trained on a portion of the data and leaves *N* data sets out of training to be used for testing. This process continues until all data sets have been used training and testing. This training approach reduces the effect of data bias.

10.4 Data Acquisition for ANN Models

ANN models can train on almost any type of data, from static databases, Excel spreadsheets or real time data from devices, the internet, sensors and so on. Analysis must consider all types of data in real time as each data source is updated. In this framework 1,073 cases were analyzed which are adequate for proper training of ANN models.

The input file for each ANN model must contain adequate, broad and diverse input variables. Often training on data variables that are highly correlated is difficult due to high correlation. Machine learning can produce interesting and robust results when there is high variety of input data and their correlation low.

10.5 Ensemble Models Case Study

In this case study, we apply deep learning methods of machine learning to predict a disease condition in patients. Prediction, for instance, enables us to apply medical preventions and interventions that can save lives.

Predicting medical conditions in patients during their hospital stay is regarded as one of the most challenging and at the same time rewarding undertakings for physicians when such predictions are timely and informative enough to allow medical intervention. During their course of care, patients frequently experience escalating health problems that lead to further medical complications.

These complications, mostly regarded as preventable (Maguire 2007), cause severe pains, injuries, disabilities and even death among patients. Several studies have suggested that complications are common with estimates of frequency ranging from 40% to 95% (Davenport, Dennis, Wellwood, Warlow 1996), and some relate poor outcomes to such complications.

In another research front, prediction has been a topic of study by engineers. To apply an engineering approach to prediction, interest in prognostics and health management (PHM) field has been growing (Pecht 2008). PHM is a discipline focused on predicting the time at which a component in a system will no longer perform as intended. When applied to medical prediction, PHM predicts when a physiological organ will fail or when a disease will occur. PHM is a relatively new field that promises to help medical prognostics (Ghavami, Kapur 2011). Among adaptive mathematical methods used for prognostics, Artificial Neural Networks (ANNs) have become popular in the last two decades as powerful prediction tools. ANNs owe their popularity to their ability to model non-linear relationships, handle adaptive learning, pattern recognition and classification—features that can be helpful in building medical predictive models.

Therefore, a computational model that can adapt to specific domains, patient demographics and geographies is desirable and useful in providing clinical predictions using available physiological data from patients under treatment. Since artificial neural networks are able to model non-linear data relationships and to adapt to new data sets, they are promising tools for this computation model.

Generally, most of the predictive methods previously proposed are based on a single model. The concept of ensemble of models (also known as committee of models) has received considerable attention in recent years. The main idea of ensemble of models is to combine the outputs of several models into a single predictor.

The concept of committee models has been used in other domains of research in the past. But only recently (Ghavami, Kapur 2011) have investigated the viability of ensemble of models in the clinical domain, trained on clinical data for medical prediction and proposed this framework to achieve higher prediction accuracy.

Patients in acute care can develop a multitude of complications ranging from pulmonary, respiratory and digestive problems due to infections. The goal of this framework is to collect such data in 1-minute intervals and study the changes in data to predict a patient's health condition using ANN models developed for each type of complication.

The challenge is to run the models in real time, once every minute as new data becomes available. The choice of time window can vary from several minutes to several hours. The ANN models dedicated to each human physiological system predict complications in each category. The physicians can apply the appropriate treatment before such complications occur or escalate. Similarly, this book explores the application of such ANN models to signal early indications as to whether treatments are being effective.

The precursor to pulmonary embolism is thrombosis, formation of blood clots inside a blood vessel obstructing the flow of blood. There are three causal elements that lead to blood clot disposition. These are known as the Virchow's triad (Virchow 1856):

- Abnormal blood flow. Abnormal blood flow is affected by narrowing of the vessels. Narrowing of the blood vessels causes turbulence that lead to formation of blood clots.
- Injuries to the vascular endothelium. Injuries to the vascular interior wall can be caused by damage to the veins arising from surgery or hypertension.
- Abnormal constitution of blood. The constituents of blood, such as proteins, water components and other elements are out of balance increasing blood thickness and propensity to form clots.

A rubric for calculating patient's risk of developing a pulmonary embolism (PE) is known as a Wells score (Wells, Anderson, et al. 1997). A Wells score provides the probability that a patient might develop a pulmonary embolism. It uses the following criteria:

- Are there clinical signs and symptoms of DVT?
- Is pulmonary embolism the top diagnosis?
- Is the heart rate over 100?
- Is the patient immobilized for at least 3 days?
- Was a surgical procedure done in the last 4 weeks?
- Was the patient previously diagnosed with PE or DVT?
- Has the patient experienced hemoptysis (coughing-up blood)?
- Does patient have malignancy with treatment within 6 months or is palliative?

The goal of predicting DVT is to use data measurements of the pre-cursors or risk factors to provide predictions about blood clot formations. The research hypothesis is that you can predict DVT in advance using data about patient's clinical data.

Finally, deep vein thrombosis (DVT) is a condition that often occurs with patients with long periods of rest in hospitals. A DVT is a blood clot that forms in a vein deep in the body, often in the lower leg or thigh. A blood clot in a deep vein can break off and travel through the blood stream. The loose blood clot is called an *embolus*. When the clot reaches the lungs and blocks blood flow, the condition is called a pulmonary embolism (PE).

When a PE is severe, it causes lungs to collapse and leads to heart failure. One in every hundred people who develop DVT dies. According to some estimates, more than 900,000 Americans develop DVT each year and 500,000 of them develop PE with 30% of those cases being fatal. About two-thirds of all DVT events are related to hospitalization. The National Quality Forum (NQF) in its 2006 update reports that DVT is the third most common cause of hospital-related deaths in the US and the most common preventable cause of hospital death.

The data collected in this case study is typically gathered over several days. The data set includes a wide range of qualitative and quantitative clinical test results and reports.

According to StopDVT.org (2011), the risk factors for DVT/PE are the following:

- Age: over 40 years
- Already had blood clots
- Family history of blood clots
- Suffering from or had treatment for cancer
- Certain blood diseases
- Being treated for heart failure and circulation problems
- Experienced recent surgery in particular in the hips or knees
- Have inherited clotting tendency
- Those who are very tall

DVT is also common among women who are:

- Pregnant
- Recently had a baby
- Taking contraceptive pills
- On hormone replacement therapy (HRT)

The initial study trained on a sample of over a thousand patients of which 225 (approximately 21%) had developed DVT/PE. Three different off-the-shelf artificial neural network tools for this study were evaluated using this data set. To perform the algorithm training, a software package was used that offered supervised learning, the ability to perform K-fold cross validation, and extensive post-test accuracy and cross-validation results after each training session.

The input data consisted of 24 different dimensions based on patient demographic and clinical elements such as: AGE, WEIGHT, GENDER, encounter type (Outpatient, Inpatient), length of stay, Stay over 48HOURS (TRUE or FALSE), ICU vs. acute care patient, BMI (BioMass index) level, Blood measures (platelets count, RBC, hematocrit, hemoglobin), other blood related values obtained from lab test results, the international normalized ratio (INR), an indicator of coagulation (INR of 2–3 is preferred but varies by patient), glucose levels, and related test results, the DVT/PE result (1 for positive, 0 for negative). This is the target variable labeled data that we can use for supervised learning. For the specific case study, sample data from the patient's electronic medical record systems was collected, and then completely anonymized so there was no identifiable information.

To build an ANN model, a typical neural tool undergoes four stages of data analysis and training:

1. Data set manager. In this stage, the scope of data fields for the model are set.
2. Train. The model is trained to identify its internal neural network weights.
3. Test. This stage determines how robust the model is (given the set of data).
4. Predict. In this stage, the trained model is applied to a new patient data set for prediction.

Once the model was trained on historical patient data, the trained model was applied to a new set of data for new patients during their stay in the hospital. The model has the ability to predict each patient's propensity to develop DVT/PE. The predictions are denoted by a '1' to denote patient is at risk of developing DVT/PE or '0' indicating that the DVT/PE risks are very low for the patient. When the prediction indicates risks of developing DVT/PE, then certain interventions such as medications and physical means are prescribed by the physician to the patient.

ANN models are known for their resilience when there is missing data. The model can fill-in the fields with missing data items. The same model can be setup to run periodically every few minutes (or every hour) for several days on the same patient but on new datasets as they're generated from the electronic medical record.

Furthermore, the model provides the input weights that it applies to compute predictions. Knowing the weights is helpful in at least two ways, including:

1. It can point to the importance of certain input variables over other variables and improves our understanding of which factors contribute to DVT/PE the most.
2. It can aid in developing hypotheses for further studies that enhance evidence-based medicine, in particular studies that determine which intervention methods have been most successful in preventing DVT/PE.

In the training of the neural net, three factors were considered: error calculation, topology selection and prevention of over-training. Error measures were computed as mean squared error over all the training cases; in other words, the mean squared difference between the correct answer and the answer given by the net. Through classification, the result is more than one output for each training case (one output corresponding to each dependent category). The tool allows computing the mean squared error over all the outputs for all the training cases in comparison to the desired output values.

The topology is determined based on the best net configuration that produces the best training result. A typical network consists of a single hidden layer. The model automatically adds a number of neurons in each layer and additional hidden layers as necessary to determine which topology learns the relationship between the independent variables and the dependent variable (response) the best (by having the lowest error). By default, the model uses two to six hidden layers. Larger models could take several hours to train. But, once the model is trained, predictions can be computed in a few seconds. Most models can be trained in two hidden layers.

Overtraining occurs when the number of iterations increases beyond the initial training such that the model's synaptic weights and topology match the problem specifically and the model is no longer generalizable to other datasets, namely the model does not apply to cases not included in the training. One approach to avoid "over-training" is the test-while-training method. In this approach the model is tested immediately after every iteration of training, then the error gets measured. If the error starts to grow, it's an indication that the researcher is starting to over-train the model.

There are four steps for building and running an ANN model.

- Step1: Define and manage the input layer data. Define the data types, independent variables and the dependent variable.
- Step 2: Train the model on a sample of cases (typically 100 cases are sufficient for training, but more cases will help reduce the percentage of bad predictions)
- Step 3: Test the model using the same set of training cases plus additional cases.
- Step 4: Run the model. Observe the error rate and percent of bad predictions.

I used a neural network software package called NeuroSolutions for this case study. A typical output screen of model training, cross validation and testing along with results are presented in Figure 10.1 as an example.

Note the mix of numeric and categorical independent variables. The dependent variable is the DVT/PE Result column. There were 1,073 independent patient cases in the data set, of which there were 225 confirmed positive cases of DVT.

The algorithm trained on 1,073 cases. The data for the purpose of this case study was simulated. Then it was tested for accuracy on an N-Leave-Out method, using 2% of data cases in each iteration for cross validation. The results of one model are shown in Figure 10.1 only as an illustration of typical output.

A typical output of cross validation shows a confusion matrix, ROC curve and other calculations including sensitivity and specificity for a range of thresholds.

Four different algorithms were used to train four independent models on the same data set. The data was randomized by the ANN tool in the first step. In the second step, a Greedy Search ANN algorithm was used to identify the significant input variables. The number of input variables was reduced from 29 to 12 according to the selection made by the first stage ANN model.

Threshold=0.3		
Output/Desired	dvtPeResult(0)	dvtPeResult(1)
dvtPeResult(0)	683	157
dvtPeResult(1)	165	68

Performance	*dvtPeResult(0)*	*dvtPeResult(1)*
MSE	0.094149096	0.598103294
NMSE	#N/A	#N/A
MAE	0.223335797	0.740626924
Min Abs Error	0.001326899	0.031017633
Max Abs Error	1.055555556	1.055555556
r	#N/A	#N/A
Percent Correct	80.54245283	30.22222222

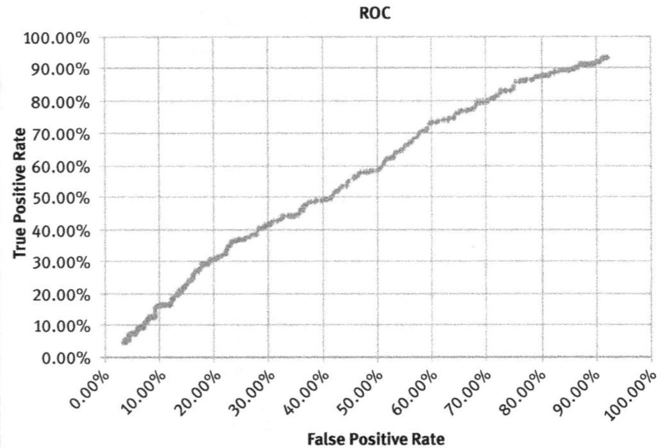

ROC Detection Threshold	*Total Detections*	*True Positive (TP)*	*False Positive (FP)*	*True Negative (TN)*	*False Negative (TN)*	*Detected as Positive ((TP+FP)/(TP+FP+TN+FN))*	*False Positive Rate (FP/(FP+TN))*	*True Positive Rate (TP/(TP+FN))*	*False Discovery Rate (FP/(FP+TP))*
0.001	991	210	781	67	15	92.36%	92.10%	93.33%	78.81%
0.002	991	210	781	67	15	92.36%	92.10%	93.33%	78.81%
0.003	989	210	779	69	15	92.17%	91.86%	93.33%	78.77%
0.004	988	210	778	70	15	92.08%	91.75%	93.33%	78.74%
0.005	985	210	775	73	15	91.80%	91.39%	93.33%	78.68%
0.006	985	210	775	73	15	91.80%	91.39%	93.33%	78.68%
0.007	985	210	775	73	15	91.80%	91.39%	93.33%	78.68%
0.008	985	210	775	73	15	91.80%	91.39%	93.33%	78.68%
0.009	984	209	775	73	16	91.71%	91.39%	92.89%	78.76%
0.01	984	209	775	73	16	91.71%	91.39%	92.89%	78.76%
0.011	984	209	775	73	16	91.71%	91.39%	92.89%	78.76%
0.012	983	209	774	74	16	91.61%	91.27%	92.89%	78.74%
0.013	983	209	774	74	16	91.61%	91.27%	92.89%	78.74%
0.014	983	209	774	74	16	91.61%	91.27%	92.89%	78.74%
0.015	982	209	773	75	16	91.52%	91.16%	92.89%	78.72%
0.016	980	209	771	77	16	91.33%	90.92%	92.89%	78.67%
0.017	979	208	771	77	17	91.24%	90.92%	92.44%	78.75%

Figure 10.1: Output screen from NeuroSolutions after training on 1,073 patient cases.

In the case of multi-layer perceptron (MLP) algorithm, the search for the optimum training level led to finding the optimum number of layers. The tool was configured to find the optimum number of layers as it constructed several multi-layer networks and compared the mean square error of the MLP networks. The model configurations consisted of 2-node, 3-node, 4-node, 5-node and 6-node arrangements. The tool selected the optimum number of layers that provided the lowest MSE. The largest number of iterations occurred with the 6-node model with 150 iterations. The model completed training and running in 40 minutes on a dual core Intel processor desktop computer (2.8GHz CPU speed).

Training each model required a separate training, cross-validation and testing process. The models consumed several "epochs" to complete the training. An epoch is a representation of an entire training set in neural networks, namely the number of iterations of training required for the model to reach its global optimum solution. The model training stops when the change in mean square error (MSE) reaches a small threshold defined by the user. Table 10.2 shows the number of epochs that each model took for training. This table compares the relative efficiency of each neural network model.

Table 10.2: Computational resources consumed by each model.

Model	Epochs
MLP- LM	20
GFN-LM	20
SVM	150
PNN	3

A sensitivity analysis was performed to determine the significance of input variables. This testing process provides a measure of the relative importance among the inputs of the neural model and illustrates how the model output varies in response to variation of an input.

Sensitivity analysis works by taking an input and varying it between its mean, ± a (user-defined) number of standard deviations, while all other inputs are fixed at their respective means. The network output is computed for a user-defined number of steps above and below the mean. This process is repeated for each input and the impact on output is recorded at each step.

An alternate variation of this process is to vary the input of interest between its minimum value and its maximum value. This option is especially useful for binary inputs or inputs which have a non-Gaussian distribution.

The result of sensitivity analysis is shown in Figure 10.2. This figure shows the relative strength of weights of the input variables. Of these variables, minimum and

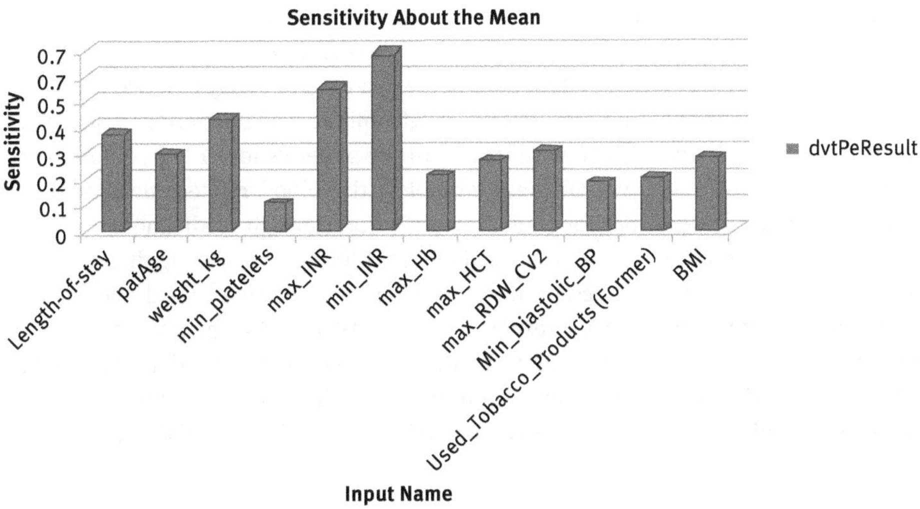

Figure 10.2: Strength of input variables toward classification.

maximum INR, followed by patient weight, length of stay, maximum RDW-CV2, and patient age were most significant input variables toward classification of patients.

The results point to several variables as being significant in predicting DVT/PE. The sensitivity analysis provided the following ranking of the input variables based on their significance toward patient classification:

1. Minimum INR
2. Maximum INR
3. Weight
4. Length of stay
5. Maximum RDW-CV2
6. Patient age

Sensitivity analysis performed using the software package reveals the degree that one variable impacts the output. It does not determine if the changes in input are positively or negatively correlated. A clinical interpretation of the sensitivity analysis can be given as follows: The INR values explain the patient's blood characteristics. It makes sense that sensitivity analysis has pinpointed minimum and maximum INR as significant inputs to classification. INR, a standard measure of blood clotting characteristics indicates the blood tendency to form clots.

A lower INR correlates with higher chance of blood clot formation. The significance of RDW-CV2 can be explained as it is a property of red blood cells that correspond to the width of red blood cells. Wider blood cells are more likely to be caught in capillaries, blocking blood flow.

Length of stay is significant as it's empirically observed that the longer the patient's stay in the hospital is correlated with increase in frequency of DVT, but only for a certain length of stay. The relationship between length of stay and occurrence of DVT follows a quadratic equation. Patient weight might point to certain underlying patient characteristics which could be related to the patient's lower levels of mobility or other factors yet unknown. Of these variables, the blood related measurements point to possible opportunities for clinical intervention through medication.

But, the benefit of sensitivity analysis is that it highlights those significant variables which can be further studied in future research to determine causal relevance to a particular disease. Unlike prior research such as Well's CPR method that choose input variables in a somewhat arbitrary method, neural network models constructed in this framework can reveal a data-driven approach through sensitivity analysis to identify the significant input variables among a large list of input variables.

Appendices

Appendix A: Prognostics Methods

Prognostics models can be classified into three general types (Eklund 2009; Hines 2009; Peysson et al. 2009). Type I is reliability based. It applies the traditional time to failure analysis by tracking a population of failures and uses statistical methods for the estimation of reliability. Some typical life distributions that are used in this type of prognostics include Weibull, exponential and normal distributions. Type I prognostic methods do not incorporate the real time monitoring of operating conditions or environmental conditions. For example, a system that has operated under harsher environments is likely to fail faster than a system based on past environmental conditions or past data.

The Weibull model is frequently used in type I methods because it offers flexible distributions for a variety of failure rate profiles. The two parameter Weibull model uses a shape parameter $\boldsymbol{\beta}$, and a characteristic life parameter $\boldsymbol{\theta}$. The result at time $\boldsymbol{t}$ is:

$$\lambda(t) = \frac{\beta}{\theta}\left(\frac{t}{\theta}\right)^{\beta-1} \tag{98}$$

Type I prognostic methods do not incorporate the real time, operating conditions or environment. For example, consider the life expectancy of a computer disk drive. The disk drive is known to have a failure distribution of 20,000 hours and standard deviation of 5,000 hours. A disadvantage of the type I approach is that it does not consider the operating condition of the system. In addition, type I prognostics offer an average for failure rate but specific failure predictions are preferable. For example, a system that has operated under harsher environments is likely to fail faster than the mean time to failure (MTTF) for that system.

Type II methods are also known as the stressor-based approaches that consider the operational and environmental condition data. Type II methods can be used if the condition data are measurable and correlated to the system degradation. This approach includes methods such as shock models and the use of traditional Markov methods. While this type of analysis is superior to Type I methods, it still lacks the unit-to-unit variance. This type considers the failures of a system in its operating environment to provide an average remaining life of a component. Some of the environmental data might include temperature, vibration, humidity and load. As an example, the proportional hazard model is a type II prognostic model. Knowing the causes, you can predict reliability of a system. The simplest model in this approach is the *regression model*: given the operating and environmental conditions, you can predict the system failure and remaining useful life by a regression equation:

$$Failure\ rate = \beta_0 + \beta_1 \times Cause_1 + \beta_2 \times Cause_2 + \cdots + \beta_n \times Cause_n \tag{99}$$

https://doi.org/10.1515/9781547401567-012

The other commonly used Type II model is the *proportional hazards model* (PHM). This model takes the environmental conditions (termed z_j) into account to modify the baseline hazard rate $\lambda_0(t)$ to produce a new hazard rate:

$$\lambda(t;z) = \lambda_0(t)e^{\left(\sum_{j=1}^{q}\beta_j z_j\right)} \tag{100}$$

The term z_j is a multiplicative factor, an explanatory variable or covariance that explains the effect on failure rate. The parameter $\lambda_0(t)$ is an arbitrary baseline hazard function and β_j is a model parameter (Eklund 2009).

As an example, the proportional hazard model is a type II prognostic model. Using the disk drive example, you can determine the expected failure rate if the disk drive's total operating hours to date and the disk drive's prior operating condition such as historical temperatures and number of disk accesses are known. These models are usually cause-and-effect based.

Type III prognostic methods are condition-based; namely they characterize the lifetime of a system in operation in its specific environment. They estimate the remaining life of a specific component or the entire system. Among methods used in Type III prognostics are the general path model (GPM), neural network models, expert systems, fuzzy rule-based systems, and multi-state analysis. Another example of type III model is the cumulative damage model.

One of the prognostic models that can gather and learn failure parameter data is artificial neural networks (ANN) and is discussed next in this chapter. As sensors become smaller and smarter, the proliferation of sensors implies increasing volume of data that can be processed for prognostics. ANNs are ideal constructs for medical prognostics because they can model feed-forward systems, compute non-linear relationships and analyze very large numbers of input data. In fact, given their parallel architecture, ANNs can function or even substitute for missing data. They are able to learn the inherent rules in a given system, can maintain long term memory and discern patterns even in noisy and changing environments. Because of these characteristics, ANNs are increasingly selected for prognostics studies and this is a reason that I've selected ANNs for this framework.

The cumulative damage model tracks the irreversible accumulation of damage in systems or components. The statistical cumulative damage model considers the number of possible damage states and a transition matrix (for representing a multi-state Markov chain) provides damage predictions for multiple cyclical loads.

Another type III method is the *shock model*. This approach is used to predict the RUL for a system subject to randomly arriving shocks. The shocks deliver certain damage of random magnitude to the system. These models are continuous in time and consequently, the degradation measures are also continuous. Shock models are estimated from historical failure data. They are similar to the Markov chain model, except that the time between shocks and the shock magnitudes are continuous and random variables.

GPM models were proposed in 1993 as a statistical method for estimating a time-to-failure distribution using degradation measures. GPM models assume that the degradation of a system is a function of time, duty cycle or some other measure. The model extrapolates a degradation function to predict RUL. GPM makes two assumptions: a) Each individual device (or system) has a unique degradation signal and b) The failure occurs at a critical threshold. This model starts with a parametric model to the exemplar degradation paths. It then computes the mean and covariance values to explain individual random parameters. It can use Bayesian probability functions to modify the posterior RUL values from a priori data. It extrapolates the critical failure threshold to estimate RUL.

The reliability and survival analysis techniques, both parametric and non-parametric methods are noteworthy of discussion. These methods as will be discussed later can be applied to prognostics. Finally, there are artificial neural network (ANN) models that are used for this framework.

Reliability and Predictive Analytics

Reliability can be defined as the probability that a product will perform its intended function satisfactorily for its intended life, when operating under specified conditions (Kapur 2010). A clinical definition can be derived from this technical description; Reliability is the probability that a patient will not develop certain medical complications during the length of stay under medical care of the care provider(s). Reliability is measured by several indicators such as *mean time between failure* (MTBF), *failure rate* and *percentiles of life*. Each measurement can be computed from corresponding equations that are derived from empirical and statistical distribution functions.

In general, reliability at time t, is shown as *R(t)*. Failures are measured by *f(t)*, the probability density function for the time of failure, a random variable *T*. The cumulative distribution function for the random variable *T* is shown by *F(t)*. Thus, you can write the following expressions to define *F(t)*, *f(t)* and *R(t)*:

$$F(t) = P\ [T \leq t] = \int_0^t f(\tau)d\tau \tag{101}$$

$$f(t) = \frac{dF(t)}{dt} \tag{102}$$

$$R(t) = 1 - F(t) \tag{103}$$

Additional treatment of reliability can be found from the text by Kapur and Lamberson (Kapur, Lamberson 1977).

Appendix B: A Neural Network Example

Consider the following classification problem as described in section 8.2. Suppose we're considering classifying patients by only four input variables, glucose (G), body mass (M), systolic blood pressure (S) and platelet count (P):

Values for G, M, S, and P for past patients are given for the model to train on, as listed below. The first step in classification is to normalize the input data to unit length. Normalizing to unit length implies that the sum of squares of values in a given data set are equal to 1. This technique was explained in Section 4.2.

The classification problem is defined as follows. There are two sets of data vectors: one set of data vectors belongs to the TRUE set (Patients with DVT) and another set belongs to the FALSE set (No DVT cases). A new patient with the normalized data is introduced and the goal is to classify that patient with the normalized values of: [0.75, 0.32, 0.60, 0.21].

The previously classified data sets and their corresponding new normalized values are computed as follows:

[117, 194, 140, 276], DVT = TRUE, normalized: [0.31, 0.51, 0.37, 0.72]
[120, 164, 213, 315], DVT = TRUE, normalized: [0.27, 0.38, 0.49, 0.73]
[115, 145, 170, 288], DVT = TRUE, normalized: [0.30, 0.38, 0.44, 0.75]
[122, 165, 155, 290], DVT = TRUE, normalized: [0.31, 0.43, 0.40, 0.75]

For patients with no DVT outcome:

[122, 144, 110, 236], DVT = FALSE, normalized: [0.38, 0.45, 0.34, 0.73]
[140, 154, 153, 176], DVT = FALSE, normalized: [0.45, 0.49, 0.49, 0.56]
[145, 135, 130, 218], DVT = FALSE, normalized: [0.45, 0.42, 0.40, 0.68]
[132, 155, 115, 190], DVT = FALSE, normalized: [0.44, 0.51, 0.38, 0.63]

The following kernel for computing PNN pattern and summation based on derivations from Equations (39) and (41):

$$z_A = f_A(x) = \sum_{i=1}^{N_k} e^{\frac{(x^t w_{ki} - 1)}{\sigma^2}}$$

The new patient with the normalized data set of [0.75, 0.32, 0.60, 0.21] is to be classified. In this example, it's assumed that σ is equal to 1.0 for sake of simplifying calculations. Next, it's possible to compute $f_A(x)$ for each data set:

$$.31^{*}.75 + .51^{*}.32 + .37^{*}.60 + .72^{*}.21 - 1.0 = -0.239 \quad \exp(-0.239) = 0.787$$

$$\exp(-0.220) = 0.803$$

$$\exp(-0.228) = 0.796$$

$$\exp(-0.231) = 0.794$$

Sum1 = 3.180

Similarly for the second class, it's possible to compute:

$$.38^{*}.75 + .45^{*}.32 + .34^{*}.60 + .73^{*}.21 - 1.0 = -0.213 \quad \exp(-0.213) = 0.808$$

$$\exp(-0.095) = 0.910$$

$$\exp(-0.144) = 0.866$$

$$\exp(-0.145) = 0.865$$

Sum2 = 3.449

Since Sum2 > Sum1, it implies that the data points for this patient are closer to a FALSE classification as the sum of values associated with the FALSE class is higher. Thus this patient belongs to the FALSE classification and is predicted to have DVT=FALSE (i.e. prognostics for DVT is negative).

Appendix C: Back Propagation Algorithm Derivation

In general, there are two types of learning processes: the *batch mode* and the *sequential mode*. One of the conditions for these mathematical calculations is that the function $\varphi_j(v_j(n))$ must be continuous and differentiable so you can obtain $\varphi_j{}'(v_j(n))$. Also the choice of η as the learning rate is important to be set at the appropriate value. The backward propagation is an approximation to the steepest descent algorithm. If the researcher uses a small η there is risk of getting unstable results. Rumelhart (1986) suggested adding a momentum value to the learning rate η according to this algorithm:

Case1: If $\frac{\partial E(n)}{\partial w_{ji}(n)}$ has the same sign for all n, then one can say that $|\Delta w_{kj}|$ grows in magnitude. This represents the case of an accelerated descent.

Case2: If in contrast $\frac{\partial E(n)}{\partial w_{ji}(n)}$ alternates its sign in every iteration, then $|\Delta w_{kj}|$ is small in magnitude and more controllable than the first case. The momentum term is significant because it accelerates learning, but also helps to avoid local optima.

The back propagation algorithm employs the following eight steps. This is adapted from Zurada (1997) and Sengupta (2009):

Given *p* training pairs shown by: $\{z_1,\ d_1,\ z_2,\ d_2,\ \ldots,\ z_P,\ d_P\}$,
where z_i is a $(I\ X\ 1)$, d_i is $(K\ X\ 1)$, and $i = 1, 2, \ldots, P$.

Step 1: Choose $> 0,\ E_{max}$. Weights W and V are initialized at small random values. W is $(K\ X\ J)$, V is $(J\ X\ 1)$. Initialize q, p and E:

$$q \leftarrow 1,\ p \leftarrow 1,\ E \leftarrow 0$$

Step 2: Start training. Input is presented and the layers' outputs are computed using

$$\Delta w_i = cf\left(w_i^t\ x\right)x:$$

$$z \leftarrow z_p,\ d \leftarrow d_p$$

$y_j \leftarrow f\left(V_j^t z\right),\ for\ j = 1, \ldots, J$ Where V_j is a column vector and is the *j*th row of V, and

$o_k \leftarrow f(W_k^t y),\ for\ k = 1, \ldots, K$ Where W_k is a column vector and is the *k*th row of W.

Step 3: Compute the *error value* by comparing the desired output versus network output:

$$E \leftarrow E + \frac{1}{2}\ (d_k - o_k)^2 \quad for \quad k = 1, \ldots, K.$$

Step 4: Compute error signal vectors δ_o and δ_y of both layers. Vector δ_o is $(K\ X\ 1)$ and δ_y is $(J\ X\ 1)$. The error signal terms of the output layer in this step are:

$$\delta_{ok} = \frac{1}{2}\ \left[(d_k - o_k)\left(1 - \delta_k^2\right)\right] \quad for \quad k = 1, \ldots, K$$

The error signal terms of the hidden layer in this step are computed by:

$$\delta_{yj} = \frac{1}{2}\left(1 - y_j^2\right)\sum_{k=1}^{K} \delta_{ok} w_{kj}, \quad for \quad j{=}1, \ldots, J$$

Step 5: Adjust the output layer weights by:

$$w_{kj} = w_{kj} + \ \eta\, \delta_{ok} y_j \quad for \quad k = 1, \ldots, K \quad and \quad j = 1, \ldots, J$$

Step 6: Adjust the hidden layer weights:

$$V_{ji} = V_{ji} + \eta\, \delta_{yj} z_i \quad for \quad i = 1, \ldots, I \quad and \quad j = 1, \ldots, J$$

Step 7: If $p < P$ then increment p and q: $q \leftarrow\ q + 1,\ p \leftarrow\ p + 1$, go to step 2, otherwise go to step 8.

Step 8: The training cycle is completed. If $E <\ E_{max}$ the training session is finished. The weights are defined by W, V, q and E. If $E >\ E_{max}$ then $p \leftarrow\ 1,\ E \leftarrow\ 0$ and initiate a new training cycle by going to step 2.

Most recently, practitioners prefer other algorithms such as the *conjugate gradient descent and simulated annealing* (Masters 1995) to backward propagation. These algorithms offer faster convergence to learning. Conjugate gradient descent is a deterministic optimization method that attempts to find the local minimum of a function. Simulated annealing is designed to ensure that the training algorithm overcomes getting trapped in local minima.

Appendix D: The Oracle Program

The oracle program is the overseer module that selects the ensemble with the highest value of AUC. The oracle program was written in R, an open source statistical package (Wang 2012) and was tested on the 1,073 data sets and all ensembles. The source code is listed below. The program computes the AUC for each ANN algorithm and ensemble compared to the actual truth (presence or absence of disease) for prior data. The first column is the actual truth and the subsequent columns are the data for ANN algorithms and ensembles.

```
########################################
###### R code for computing AUC    ######
###### Peter Ghavami - Adapted     ######
######     From Wang 2012          `######
########################################

### Load library packages ###

install.packages("verification")
library(verification)

### Import the input data file ###

## select the data file in the pop out window
data<-read.csv(file.choose(),header=T) # Import data from csv file

### Purpose: write a R function to compute AUC ###
### function name: getAUC
### function input:
###     D: true disease status
###     T: continuous test results of ANN and Ensembles
### function output: AUC value for this test

getAUC<-function(D, T){
roc.area(D, T)$A
}                              # Call the getAUC function.

### compute the AUC values for each test in the data

### result for MLP
getAUC(data[,1],data[,2]) #MLP data is in the 2nd column

### result for PNN
getAUC(data[,1],data[,6])  #PNN data is in the 6th column
```

```
### result for GFF
getAUC(data[,1],data[,10])  #GFF data is in the 10th column

### result for SVM
getAUC(data[,1],data[,14])  #SVM data is in the 14th column

### result for Ensemble1
getAUC(data[,1],data[,18])  #Ensemble1 data is in the 18th column

### result for Ensemble2
getAUC(data[,1],data[,22])  #Ensemble2 data is in the 22nd column

### result for Ensemble3
getAUC(data[,1],data[,26]) #Ensemble3 data is in the 22nd column

### result for Ensemble4
getAUC(data[,1],data[,30])  #Ensemble4 data is in the 30th column

### result for Ensemble5
getAUC(data[,1],data[,34])  #Ensemble5 data is in the 34th column
```

References

Aamodt, A., Plaza, E., "Case-Based Reasoning: Foundational Issues, Methodological Variations, and System Approaches," *AI Communications*, Vol. 7, No. 1, March 1994.

Adams, J. B., Wert, Y., "Logistic and Neural Network Models for Predicting a Hospital Admission," *Journal of Applied Statistics*, Vol. 32, No. 8, 861–869, 2005.

Allison, P. D., *Survival Analysis Using the SAS System*, SAS Institute publication, Cary, NC, 1995.

AMA 2010. CPT 2011 Professional Edition, Michelle Abraham, *American Medical Association*, American Medical Association Press, Oct. 20, 2010.

Arthi, K., Tamilarasi, A., "Prediction of Autistic Disorder Using Neuro Fuzzy Systems by Applying ANN Technique," *International Journal of Developmental Neuroscience*, Vol. 26, 699–704, 2008.

Baxt, W. G., "Use of an Artificial Neural Network for the Diagnosis of Myocardial Infarction," *Annals of Internal Medicine*, Dec, 1, Vol. 115, No. 11, 843–848, 1991.

Bewick, V., Cheek, L., Ball, J., "Statistics Review 13: Receiver Operating Characteristic Curves," *Critical Care*, Vol. 8, No. 6, December 2004.

Blount, M., Ebling, M. R., Eklund, J. M., James, A. G., McGregor, C., Percival, N., Smith, K. P., Sow, D., "Real-time Analysis for Intensive Care. Development and Deployment of the Artemis Analytic System," *IEEE Engineering in Medicine and Biology Magazine*, March/April 2010.

Bourdes, V., Ferrieres, J., Amar, J., Amelineau, E., Bonnevay, S., Berlion, M., Danchin, N., "Prediction of Persistence of Combined Evidence-based Cardiovascular Medications in Patients with Acute Coronary Syndrome after Hospital Discharge Using Neural Networks," *Medical & Biological Engineering Computing*, Vol. 49, 947–955, 2011.

Bottaci, L., Drew, P. J., Hartley, J. E., Hadfield, M. B., Farouk, R., Lee, P. WR., Macintyre, I. MC., Duthie, G. S., Monson, J. RT, "Artificial Neural Networks Applied to Outcome Prediction for Colorectal Cancer Patients in Separate Institutions," *The Lancet*, Vol. 350, No. 9076, 469–472, Aug. 16, 1997.

Breiman, L., Friedman, J. H., Olshen, R. A., Stone, C. J., *Classification and Regression Trees.* Monterey, CA: Wadsworth & Brooks/Cole Advanced Books & Software, 1984.

Brown, S.W., Strong, V., "The Use of Seizure-alert Dogs," *Seizure*, Vol. 10, 39–41, 2001.

Center for Evidence Based Medicine website, EBM Tools, http://www.cebm.net/index.aspx?o=1023, accessed March 22, 2012.

Coble, J., Hines, J. W., "Identifying Optimal Prognostic Parameters from Data: A Genetic Algorithms Approach," Annual Conference of the Prognostics and Health Management Society, 2009.

Coble, J., Hines, J. W., Fusing Data Sources for Optimal Prognostic Parameter Selection, Sixth American Nuclear Society International Topical Meeting on Nuclear Plant Instrumentation, Control, and Human-Machine Interface Technologies NPIC & HMIT 2009, Knoxville, Tennessee, April 5–9, 2009.

Collett, D., *Modeling Survival Data in Medical Research*. London: Chapman & Hall, 1994.

Daley, M., Narayanan, N., Leffler, C. W., "Model-derived Assessment of Cerebrovascular Resistance and Cerebral Blood Flow Following Traumatic Brain Injury," *Experimental Biology and Medicine*, Vol. 235, April 2010.

Davenport, R.J., Dennis, M.S., Wellwood, I., Warlow, C., "Complications After Acute Stroke," *Stroke*, Vol. 27, 415–420, 1996.

Dayhoff, J. E., DeLeo, J. M., "Artificial Neural Networks, Opening the Black Box," *Cancer*, Vol. 19, 1615–1635. Presented at the Conference on Prognostic Factors and Staging in Cancer Management: Contributions of Artificial Neural Networks and Other Statistical Methods, 2001.

Delen, D., "Analysis of Cancer Data: A Data Mining Approach," *Expert Systems*, February 2009, Vol. 26, No. 1.

https://doi.org/10.1515/9781547401567-013

Dictionary.com www.Dictionary.com, online, an IAC company, Accessed, Jan. 2012. Doyle, J., Francis, B., Tannenbaum, A., *Feedback Control Theory*, Macmillan Publishing Co, 1990.

Dybowski, R., Gant, V., Weller, P., and Chang, R., "Prediction of Outcome in Critically ill Patients Using Artificial Neural Network Synthesised by Genetic Algorithm," *The Lancet*, Vol. 347, No. 9009, 1146–1150April 27, 1996.

Eklund, N. H. W., "Prognostics and Health Management—Part 1: Data Driven Anomaly Detection & Diagnosis," Annual Conference of the Prognostics and Health Management Society, Diagnostics Tutorials, 2009.

Floyd, C. E., Lo, J. Y., Yun, A. J., Sullivan, D. C., Kornguth, P. J., "Prediction of Breast Cancer Malignancy Using an Artificial Neural Network," *Cancer*, Vol. 74, No. 11, Dec. 1, 1994.

Fuller, R. L., McCullough, E. C., Bao, M. Z., Averill, R. F., "Estimating the Costs of Potentially Preventable Hospital Acquired Complications, " *Healthcare Financing Review*, Vol. 30, No. 4, Summer 2009.

Gao, E., Young W., Ornstein, E., Pile-Spellman, J., Qiyuan, M., "A Theoretical Model of Cerebral Hemodynamics: Application to the Study of Arteriovenous Malformations," *Journal of Cerebral Blood Flow and Metabolism*, 17, 905–918, 1997.

Ghavami, P., *Clinical Intelligence: The Big Data Analytics Revolution in Healthcare – A Framework for Clinical and Business Intelligence*, Amazon Publishing, 2014.

Ghavami, P., Kapur, K., "Prognostics & Artificial Neural Network Applications in Patient Healthcare," *Proceedings of IEEE Prognostics and Health Management Conference*, June 2011.

Graunt, J., Natural and Political Observations Made Upon the Bills of Mortality, 1665.

Hahnfeldt, P., Panigraphy, D., Folkman, J., Hlatkey, L., "Tumor Development under Angiogenic Signaling: A Dynamic Theory of Tumor Growth, Treatment Response and Postvascular Dormacy," *Cancer Research* 59, 4770–4778, 1999.

Hansen, B. and Klopfer, S.O., "Optimal Full Matching and Related Designs via Network Flows," *Journal of Computational and Graphical Statistics*. Vol. 15, No. 3, 2006.

Hardy, M., "Gaussian Function with 2-dimensional Domain," Wikipedia commons. Originally developed as "Isometric plot of a two dimensional gaussian," created by Kaushik Ghose using MATLAB, 2006.

Haykin, S., *Neural Networks, A Comprehensive Foundation*, 2nd Edition, Prentice Hall, 1998.

Hornik, K., Stinchcombe, M., White, H., Multilayer feedback networks are universal approximators, *Journal of Neural Networks*, Vol. 2(5), 359–366, 1989. Elsevier Science Ltd, Oxford, UK.

Hines W. J., "Empirical Methods for Process and Equipment Prognostics," Annual Conference of the Prognostics and Health Management Society, Prognostics Tutorials, 2009.

Hu, C., Youn, B.D., Wang, P., "Ensemble of Data-driven Prognostics Algorithms with Weight Optimization and K-Fold Cross Validation," Annual Conference of the Prognostics and Health Management (PHM) Society, Oct. 10–16, 2010, Portland, OR.

INCOSE, What is a System?, Version 2.0, *INCOSE (International Council on Systems Engineering Council) Systems Engineering Handbook*, July 2000.

Jervis, R., McGinn, T., "Evidence-based Medicine, Clinical Prediction Rules for Hospitals," *Mount Sinai Journal of Medicine*, Vol. 75, 472–477, 2008.

Kalilani, L., Atashili, J., "Measuring Additive Interaction Using Odds Ratios," *Epidemiol Perspect Innov.*, Vol. 3, 5, 2006.

Kapur, K., *Seminar on Prognostics, Dept. of Industrial & Systems Engineering, University of Washington*, Feb.–March 2010.

Kapur, K., Lamberson, L. R., *Reliability in Engineering Design*, 1977.

Kimmel, M., Axelrod, D.E., *Branching Processes in Biology*, Springer Verlag, New York, NY, 2002.

Kirton, A., Winter, A., Wirrell, E., Snead, O. C., "Seizure Response Dogs: Evaluation of a Formal Training Program, " *Epilepsy & Behavior*, Vol. 13, 499–504, 2008.

Kodell, R. L., Pearce, B. A., Baek, S., Moon, H., Ahn, H., "A Model-free Ensemble Method for Class Prediction with Application to Biomedical Decision Making," *Artificial Intelligence in Medicine*, Vol. 46, 267–276, 2009.

Kon, A., M., Plaskota, L., "Complexity of Predictive Neural Networks," International Conference on Complex Systems, May 2000.

Kwakernaak, H., Sivan, R., *Linear Optimal Control Systems*, John Wiley & Sons, 1972.

Laupacis, A., Sekar, N., Stiell, I. G., "Clinical Prediction Rules. A Review and Suggested Modifications of Methodological Standards," *JAMA*, Vol. 277, 488–494, 1997.

Ling, C. X., Huang, J., Zhang, H., "AUC: A Statistically Consistent and More Discriminating Measure than Accuracy," *International Joint Conference on Artificial Intelligence*, Vol. 18, 519–526, Lawrence Erlbaum Associates, LTD, 2003.

Ling, C. X., Huang, J., Zhang, H., "AUC: A Better Measure than Accuracy in Comparing Learning Algorithms," *Lecture Notes in Computer Science*, ISSU 2671, 329–341, Springer-Verlag, 2003.

Limaye, S. S., Mastrangelo, C. M., Zerr, D. M., Jeffries, H., "A Statistical Approach to Reduce Hospital-associated Infections," *Quality Engineering*, Vol. 20, 414–425, 2008.

Linder, R. Geier, J., Kolliker, M., "Artificial Neural Networks, Classification Trees, and Regression: Which Method for Which Customer Base?" *Database Marketing & Customer Strategy Management*, Vol. 11, No. 4, 344–356, 2004.

Lisboa, P. J., Taktak, A. F.G., "The Use of Artificial Networks in Decision Support in Cancer: A Systematic Review," *Neural Networks*, Vol. 19, No. 4, 408–415, May 2006.

Lucchetti, R., "Convexity and Well-posed Problems," *CMS Books in Mathematics*, 2006.

Macal, C., "Model Verification and Validation, The University of Chicago and Argonne National Laboratory," Workshop on "Threat Anticipation: Social Science Methods and Models," Chicago, IL, April 7–9, 2005.

Maguire, P., "The New Crackdown on Preventable Complications," *Today's Hospitalist*, October 2007.

Masters, T., *Advanced Algorithms for Neural Networks: A C++ Sourcebook*, Wiley, New York, 1995.

McGinn, T. G., Guyatt, G. H., Wyer, P. C., Naylor, C. D., Stiell, I. G., Richardson, W. S., "Users' Guide to Medical Literature," *JAMA*, Vol. 284, No. 1, 79–84; For the Evidence-based Medicine Working Group, 2000.

Merriam-Webster dictionary online, www.Merriam-webster.com/dictionary/, an Encyclopedia Britannica Company, accessed December 2011.

MIT, see http://classics.Mit.edu/Hippocrates/prognost.html, accessed Feb. 2010.

Monterola, C., Lim, M., Garcia, J., Saloma, C., "Feasibility of a Neural Network as Classifier of Undecided Respondents in a Public Opinion Survey," *International Journal of Public Opinion Research*, Vol. 14, No. 2, 2002.

NeuroDimension, Inc., Gainesville, Florida, NeuroSolutions software, Version 6.0, 2011.

NHS Casemix, "The Casemix Design Framework—2009," by Casemix Design Authority. Version 2.3, Issue Date: December 2009. The Health and Social Care Information Centre, Casemix Service.

Niu, G., Yang, B., Pecht, M., "Development of an Optimized Condition-based Maintenance System by Data Fusion and Reliability-centered Maintenance," *Reliability Engineering and System Safety*, Vol. 95, No. 7, 786–796, 2010.

O'Connor, A. M., Bennett, C.L., Stacey, D., Barry, M., Col, N. F., Eden, K.B., Entwistle, V. A., Fiset, V., "Decision Aids for People Facing Health Treatment or Screening Decisions (Review)," *The Cochrane Collaboration*, Wiley, 2009.

Ozbay, H., *Introduction to Feedback Control Theory*, CRC Press, 1999.

Park, Y., Kim, B., Chun, S., "New Knowledge Extraction Technique Using Probability for Case-based Reasoning: Application to Medical Diagnosis," *Expert Systems*, Vol. 23, No. 1, Feb. 2006.

Pecht, M., *Prognostics and Health Management of Electronics*, Wiley, 2008.

Peysson, F., Ouladsine, M., Outbib R., "Complex System Prognostics: A New Systemic Approach," Annual Conference of the Prognostics and Health Management Society, 2009.

Principe, J. C., Euliano, N. R., Lefebvre, W. C., *Neural and Adaptive Systems, Fundamentals Through Simulations*, John Wiley & Sons, 1999.

Principe, J. C., Conversations with Jose C. Principe, University of Texas, Sept. 2011.

Prodormidis, A. L., Chan, P. K., Stolfo, S. J., "Meta-learning in Distributed Data Mining Systems: Issues and Approaches," *Advances in Distributed Data Mining*, MIT Press, 2000.

Ravdin, P. M. and Clark, G. M., "A Practical Application of Neural Network Analysis for Predicting Outcome of Individual Breast Cancer Patients," *Breast Cancer Research and Treatment*, Vol. 22, No. 3, 285–293, Oct. 1992.

Rosenbaum, R., Rubin, D., "The Central Role of Propensity Score in Observational Studies for Causal Effects," *Biometrika*, Vol. 70, No. 1, 41–55, 1983.

Rumelhart, D. E., Hinton, G. E., and Williams, R. J., "Learning Representations by Back-propagating Errors," *Nature*, Vol. 323, 533–536, 1986.

Schlimmer, J. C., Granger, Jr., R. H., "Incremental Learning from Noisy Data," *Machine Learning*, Vol. 1, 317–354, Kluwer Publishers, Boston, 1986.

Sengupta, S., *Lecture series on Neural Networks and Applications by Prof. S. Sengupta*, Department of Electronics and Electrical Communication Engineering, Indian Institute of Technology, Kharagpur, source: NPTEL, http://nptel.iitm.ac.in, accessed 2009–2012.

Smye, S. W., Clayton, R. H., "Mathematical Modeling for the New Millennium: Medicine by Numbers," *Medical Engineering & Physics*, Vol. 24, 565–574, 2002.

Souter, M., *Conversations on Diagnostic Markers and Predictors*, April 7, 2011.

Spiegelman, D., Schneeweiss, S., McDermott, A., "Measurement Error Correction for Logistic Regression Models with an 'Alloyed Gold Standard'," *American Journal of Epidemiology*, Vol. 145, No. 2, 1996.

Spruance, S. L., Reid, J. E., Grace, M., Samore, M., "Hazard Ratio in Clinical Trials," *Antimicrobial Agents and Chemotherapy*, Vol. 48, No. 8, Aug. 2004.

StopDVT.org, www.StopDVT.org website, http://stopdvt.org/FAQ.aspx, accessed Oct. 2011.

Strong, V., Brown, S. W., Walker, R., "Seizure-alert Dogs-Fact or Fiction?" *Seizure*, Vol. 8, 26–65, 1999.

Swierniak, A., Kimmel, M., Smieja, J., "Mathematical Modeling as a Tool for Planning Anticancer Therapy," *European Journal of Pharmacology*, Vol. 625, 108–121, 2009.

Toll, D. B., Janssen, K. J. M., Vergouwe, Y., Moons, K. G. M., "Validation, Updating and Impact of Clinical Prediction Rules: A Review," *Journal of Clinical Epidemiology*, Vol. 61, 1085–1094, 2008.

Tsai, K., Pollock, K., Brownie, C., "Effects of Violation of Assumptions for Survival Analysis Methods in Radiotelemetry Studies," *Journal of Wildlife Management*, Vol. 63, No. 4, 1369–1375, 1999.

TU, J. V., "Advantages and Disadvantages of Using Artificial Neural Networks versus Logistic Regression for Predicting Medical Outcomes, *Journal of Clinical Epidemiology*, Vol. 49, No. 11, 1225–1231, Nov. 1996.

Uckun, S., Goebel, K. and Lucas, P. J. F., "Standardizing Research Methods for Prognostics," 2008 International Conference on Prognostics and Health Management, 2008.

Vichare, N. M., and Pecht, M., "Prognostics and Health Management of Electronics," *IEEE Transactions on Components and Packaging Technologies*, Vol 29, No. 1, March 2006.

Virchow, R., Virchow's Triad. Virchow's Triad was first formulated by the German physician Rudolf Virchow in 1856.

Wang, Z., *Conversations about Neural Network Algorithms and Accuracy Measures*. Department of Biostatistics, University of Washington, 2012.

Wang, Z., *Conversations and Collaboration for Calculating AUC using R Statistical Language.* Department of Biostatistics, University of Washington, 2012.

Webber, W. R. S., Litt, B., Wilson, K., Lesser, R. P., "Practical Detection of Epileptiform Discharges (EDs) in the EEG Using an Artificial Neural Network: A Comparison of Raw and Parameterized EEG Data." *Electroencephalography and Clinical Neurophysiology*, Vol. 91, 194–204, 1994.

Wells, P. S., Anderson, D. R., Bromanis, J., Guy, F., Mitchell, M., Gray, L., Clement, C., Robinson, K. S., Lewandowski, B., "Value of Assessment of Pretest Probability of Deep-vein Thrombosis in Clinical Management," *The Lancet*, Vol. 350, No. 9094, 1795–1798, December 20, 1997.

WHO, *Library of ICD9 and ICD10 Codes*, World Health Organization's library of International Statistical Classification of Diseases and Related Health Problems, http://www.who.int/classifications/icd/revision/en/index.html, accessed March 9, 2012.

Williamowski, B. M., Chen, Y., "Efficient Algorithm for Training Neural Networks with One Hidden Layer," *IEEE International Joint Conference on Neural Networks*, 1999.

Williams, H., Pembroke, A., "Sniffer Dogs in the Melanoma Clinic?" *Lancet*, Vol. 1, No. 8640, 734, 1989.

Wishart, D., "Symposium on Control Theory: A Survey of Control Theory," *Journal of the Royal Statistical Society, Series A*, Royal Statistical Society, 1969.

Yu, C., Liu, Z., McKenna, T., Reisner, A. T., Reifman, J., "A Method for Automatic Identification of Reliable Heart Rates Calculated from ECG and PPG Waveforms," *Journal of the American Medical Informatics Association*, Vol. 13, No. 3, May/June 2006.

Zadeh, L. A., and Desoer, C., *Linear Control Theory*, Springer-Verlag, 1963.

Zurada, J. M., *Introduction to Artificial Neural Network*, Jaico Publishing House, Second Edition, 1997.

Index

https://doi.org/10.1515/9781547401567-014

www.ingramcontent.com/pod-product-compliance
Lightning Source LLC
LaVergne TN
LVHW060032210926
848481LV00013B/3844